Edible Wild Plants of
the Prairie

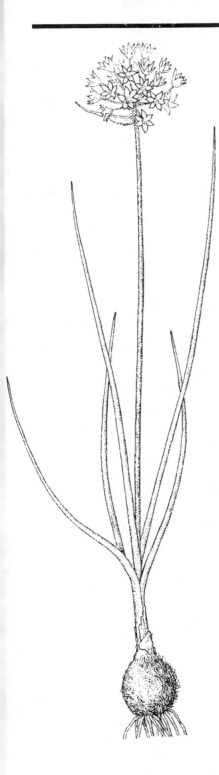

Edible
Wild Plants
of the
Prairie

An Ethnobotanical
Guide

Kelly Kindscher

Drawings by Carol Kuhn

University Press of Kansas

To

my mother

and father

A grant from the Faultless Starch/Bon Ami Company made possible the research for and writing of this book, as well as the line drawings of the plants.

The map on page 2 is reprinted from David Costello, *The Prairie World* (Minneapolis: University of Minnesota Press, 1969), by permission of the author.

Published by the University Press of Kansas (Lawrence, Kansas 66049), which was organized by the Kansas Board of Regents and is operated and funded by Emporia State University, Fort Hays State University, Kansas State University, Pittsburg State University, the University of Kansas, and Wichita State University

Library of Congress Cataloging-in-Publication Data: See p. 277
Printed in the United States of America
10 9 8 7 6

Contents

Preface

Starting on May 5, 1983, Vicky Foth and I walked for the next eighty days across Kansas and eastern Colorado, from the mouth of the Kansas River in Kansas City to the foothills of the Rocky Mountains south of Denver. We walked 690 miles, averaging about 10 miles a day, traveling parallel to the Santa Fe and Smoky Hills trails by following county roads that were usually unpaved, or going cross-country through areas of rangeland.

It was exciting to be immersed in the wildness and beauty of the region, especially when the wheat waved its bountiful green heads in the fields as we walked by, and where the native pasture grasses were greening underneath our feet while a myriad of wildflowers delicately accented the landscape. Of course there were days of sore feet, tired muscles, seemingly endless rain, and, further west, searing heat and a shortage of drinking water. At the end of most days, after carrying a forty-five pound pack for about eight hours, I would set it down at our campsite and take a light-footed walk to the top of the nearest hill. Here I would watch the birds of the prairie soar in the evening light, while lying down and resting amongst the grass and flowers.

One purpose of this walk was to broaden my perspective and to deepen my knowledge of the Prairie Bioregion. My interest in prairie plants began while I was growing up and, later, spending weekends and summers on my family's farm near Guide Rock, Nebraska, not far from a historic Pawnee Indian village along the Republican River. My father taught me about plants while we were fixing fence in the pasture that was part of the original 1871 homestead claim of my Kindscher great-grandparents. During the walk, my thoughts repeatedly turned to how life must have been for these homesteaders and for the Indians before them. Along the way, I made extensive notes on the edible plants and sampled most of them. This personal experience became the inspiration for this book.

Another result of the walk was a deeper understanding of the spirit of the prairie. This understanding was clearly expressed by Melvin Gilmore, whose work in the 1910s made the most important contribution to the ethnobotany of the region: "For me, since I have acquired from the old Indians of many tribes of this region . . . the lore of places, plants and animals, the country is alive with interest and spirit. It lives with me and talks to me. On any trip, by rail, automobile, horseback or on foot, the plants along the way, the birds that fly, and the mammals which I may chance to see, all have their story and song" (Nickel, 1971, p. 32). My hope is that the stories and songs of the plants in this book will inspire others to learn more about the prairie and its heritage.

Many people provided assistance for this project. I thank members of the Department of History at the University of Kansas—Rita Napier, Norman Saul, and Phil Paludan—who arranged my appointment as a research associate. This allowed me to use a research study in Watson Library. I am also indebted to the Interlibrary Loan Department staff for obtaining copies of many obscure books.

I am grateful to Elaine Shea and the Grassland Heritage Foundation for providing seed money and to Gordon and Cathy Beaham of the Faultless Starch/Bon Ami Company, whose concern for the environment and prairie resulted in the financial support to finish this project. Further organizational help was given by the Kansas Area Watershed (KAW) Council and the Appropriate Technology Center, which provided a computer for the project (donated by Pierce Butler of Laurel Hill Plantation). Additional computer assistance was offered by Dean Thompson, Carl Thor, and Bob Nunley.

I received helpful criticism and support during the writing from my friends Dan Bentley, Ken Lassman, Chuck Magerl, Holly Extner-Thompson, and Sandy Strand; from my mother, Charlene Kindscher; and especially from Susan Jones.

I greatly appreciate the technical assistance, advice, and other support I received from Richard Ford, University of Michigan Ethnobotanical Laboratory; Raymond DeMallie, Department of Anthropology, Indiana University; Waldo Wedel, Archaeologist Emeritus, Smithsonian Institution; Gary Nabhan, Desert Botanical Garden, Phoenix, Ariz.; Kay Young, Fontenelle Forest Nature Center, Omaha, Nebr.; Mary Adair, Archaeo-botanical Consulting, Kansas City, Mo.; William Blackmon, Department of Crop Physiology, Louisiana State University; Sue Richman, Goddard College; Dwight Platt, Bethel College; Ron McGregor and Ralph Brooks, University of Kansas Herbarium; Bill and Jan Whitney, Prairie/Plains Resource Institute; Steven Foster, Izard Ozark Natives; Wes Jackson, the Land Institute; and Vinnie McKinney, Elixir Farm.

Special thanks go to the staff of the University Press of Kansas for continued support and encouragement, and to the reviewers who were selected: E. Barrie Kavasch of Native Harvest Associates and Ted Barkley and Patricia O'Brien of Kansas State University. They provided invaluable comments and suggestions. Eileen K. Schofield, Kansas State University, wrote the botanical descriptions and edited the final manuscript. Carol Kuhn, the illustrator, has been a pleasure to work with, and her line drawings bring to life the story of each plant.

Introduction

These are the stories of native prairie plants: how they were used as food by prehistoric and historic Indians, early explorers, and travelers of the region and how they can be used today. There are many books about the edible plants of eastern North America, as well as some on western North America. This one highlights the region in between—the Prairie Bioregion.

Bioregions are geographical areas whose soft boundaries are set by nature. The Prairie Bioregion is distinguished from the neighboring Rocky Mountain and Ozark bioregions by characteristic plants, animals, water relations, climate, and geology. For a further discussion of the concept of bioregions, see *Dwellers in the Land: The Bioregional Vision* (Sale, 1985) and "On 'Bioregionalism' and 'Watershed Consciousness'" (Parsons, 1985).

The Prairie Bioregion is immense, covering over 1 million square miles. It stretches from Texas north to Saskatchewan and from the Rocky Mountains (from New Mexico to Montana) in the west to the deciduous forests of Missouri, Indiana, and Wisconsin in the east (see Map 1). It has a rich variety of grasses and forbs that are drought tolerant and need full sunlight. One obvious feature of the prairie is the relative absence of trees. In the prairie's native condition, trees were kept out by fires set by Indians or lightning, drought, and grazing by buffalo, antelope, and elk, all of which kept shrubs and trees from becoming established except in moist, protected areas.

George Catlin was an artist who made many historical paintings of the Prairie Bioregion and its native people. In 1832 near Fort Leavenworth (along the Missouri River in northeast Kansas), Catlin (1973, vol. 1, pp. 19–20) described the awesome and exciting prairie fires in a passage that paints a vivid picture of the unspoiled prairie:

The war, or hell of fires! where the grass is seven or eight feet high, as is often the case for many miles together, on the Missouri bottoms; and the flames are driven forward by the hurricanes, which often sweep over the vast prairies of this denuded country. There are many of these meadows on the Missouri, the Platte, and the Arkansas, of many miles in breadth, which are perfectly level, with a waving grass, so high, that we are obliged to stand erect in our stirrups, in order to look over its waving tops, as we are riding through it. The fire in these, before such a wind, travels at an immense and frightful rate, and often destroys, on the fleetest horses, parties of Indians, who are so unlucky as to be overtaken by it; not that it travels as fast as a horse at full speed, but that the high grass is filled with wild peavines and other impediments, which render it necessary for the rider to guide his horse in the zigzag paths of the deer and buffa-

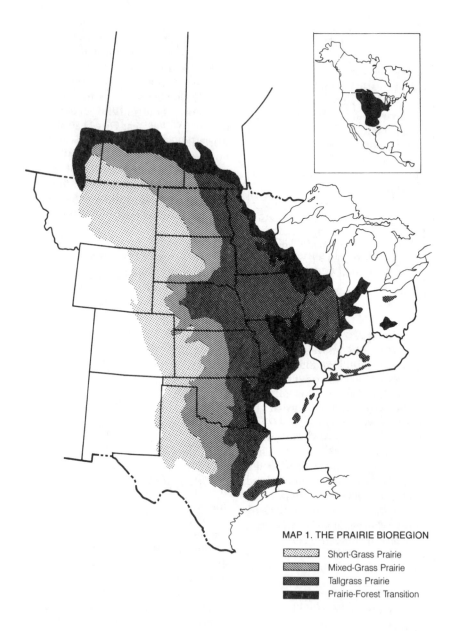

MAP 1. THE PRAIRIE BIOREGION

Short-Grass Prairie
Mixed-Grass Prairie
Tallgrass Prairie
Prairie-Forest Transition

loes, retarding his progress, until he is overtaken by the dense column of smoke that is swept before the fire—alarming the horse, which stops and stands terrified and immutable, till the burning grass which is wafted in the wind, falls about him, kindling up in a moment a thousand new fires, which are instantly wrapped in the swelling flood of smoke that is moving on like a black thunder-cloud, rolling on the earth, with lightning's glare and its thunder rumbling as it goes.

The observations of other early explorers show that some perceived the treeless prairie as lonely. John Treat Irving after traveling through eastern Kansas prairies in 1833, called them a "green waste" and later commented, "All was prairie. . . . A tree would have been a companion, a friend" (McDermott, 1955, pp. 11, 53). Charles Pruess, the cartographer on the Frémont Expedition, reported on June 12, 1842, while in present-day northeast Kansas, heading west along the Kansas River: "Eternal prairie and grass, with occasional groups of trees. . . . To me it is as if someone would prefer a book with blank pages to a good story. The ocean has, after all, its storms and icebergs, the beautiful sunrise and sunset. But the prairie? To deuce with such a life, I wish I were in Washington with my old girl" (Gudde and Gudde, 1958, p. 5).

The European woodland perspective of our dominant historical roots still seems to dictate views of the Prairie Bioregion. Public parks throughout the region are rarely prairie; most often they are nonnative species of grasses amongst trees. When people think of wildflowers, wild edible plants, or even birds, they seldom think of prairie species, even if they live in the Prairie Bioregion or near a prairie remnant. Wilderness connotes areas of virgin forest rather than virgin prairie. Because of these perceptions, it is not surprising that the prairie is still not represented in our national park system and that its uniqueness has been so little appreciated.

The Prairie Bioregion can be divided into three north-south zones: the tall-grass prairie in the east; the mixed-grass prairie in the center; and the short-grass prairie in the west. The tall-grass prairie is dominated by big bluestem (*Andropogon gerardii*), Indian grass (*Sorghastrum nutans*), switch grass (*Panicum virgatum*), and little bluestem (*Andropogon scoparius*). The soil here is rich and fertile and rainfall is adequate, so much of the area has been plowed for agricultural crops.

The short-grass prairie features buffalo grass (*Buchloë dactyloides*), blue and hairy grama grass (*Bouteloua gracilis* and *B. hirsuta*). This is where the great herds of buffalo once roamed. Because of the dry climate, the area is suited to grazing, but drought-tolerant and irrigated crops are grown.

As the name implies, the mixed-grass prairie contains a mixture of species from the other two zones. The proportions of various plants change over the years, depending on rainfall cycles. Overall, the flora of the Prairie Bioregion is quite varied. In addition to the dominant grasses, there are a wide variety of herbaceous plants (often called forbs) and woody plants.

Edible prairie plants can be divided into these three general categories—forbs (e.g., prairie turnip), woody shrubs (e.g., chokecherry), and grasses (e.g., Eastern gama grass).

The importance of wild plant foods to the Indians has been overshadowed by the romance of the adventure of hunting the large grassland herbivores: antelope, elk, and especially buffalo. Yet the gathering of these foods, a necessary seasonal activity, was as important to the native peoples as hunting. Wild plants were a significant food source to complement their meat diet, providing starches, vitamins, and minerals. Vegetable foods, including the horticultural crops (corn, beans, and squash) grown by the semi-sedentary Indians of the region and traded to the nomadic tribes, were an important—if not the most important—part of the diet of the Indians of the Prairie Bio-

region. The Omaha Indians, for example, were known to determine the route of their summer buffalo hunt not by where they would find buffalo, but by the locations of prairie turnips and other wild foods (Fletcher and LaFlesche, 1911). They made their camps at these places so that both vegetable and animal foods could be taken and processed for later use.

Elias Yanovsky, in *Food Plants of the North American Indians* (1936), listed 1,112 species of plants that were used as food sources. In *Uses of Plants by the Indians of the Missouri River Region* (1977), Melvin Gilmore reported over 150 species that were used for medicine (the major use), food, and other purposes. Because grasses dominate the vegetation of the Prairie Bioregion and all of them have edible seeds, it could be said that the majority of plants are edible. However, most grass seeds are quite small, enclosed in a tough hull, and there is very little archaeological or ethnographic evidence for the use of grass seeds, so they probably were not major sources of food.

This book documents the Indian use of 123 species of native prairie plants for food. Certainly many other plants were used as minor sources of food, and in times of scarcity, probably almost anything that could be eaten was eaten. Generally, a wide variety of plant foods were eaten by choice for a variety of reasons: they were

4

readily available, nutritious, and tasty; they relieved a monotonous meat diet; they could be eaten in the spring when other foods (game, stored foods, and any crops, for horticultural tribes) were scarce; and they helped to avert starvation when hunting failed.

Archaeological studies in the region, particularly the analysis of scattered plant remains, have given us some information on the plants that native people used as food. However, many fleshy plant materials do not preserve well in the open and exposed prairie environment. Analysis of remains found in sheltered environments in the adjacent Ozarks and desert Southwest provides further clues to possible prehistoric plant uses in the Prairie Bioregion. For example, human coprolites found in those areas contain food remains and plant pollen that are direct proof of what the people ate.

The role of humans in the dispersal of plants is often unrecognized. It is quite possible that the distribution of some edible plants within the Prairie Bioregion was expanded significantly by the native people. For example, when the Pawnee were relocated to a reservation in Oklahoma in the 1870s, they brought with them wild plums from their homelands in Nebraska.

Indian women were the primary gatherers of wild food plants; only on rare occasions did men participate. Most of the early European contacts with the Indians were made by men—explorers, trappers, traders—who viewed Indian cultures as based on the hunting activities of men. Perhaps this is one reason why the importance of plant gathering by Plains Indian women was underestimated.

By the time the prairies were being settled by whites, most of the Plains Indians had been removed to, and isolated on, reservations. There was almost no interaction between Indian women, who knew the wild food plants, and women settlers, whose responsibilities included gardening and food preparation. Consequently, except for wild onions and some fruits (chokecherries, plums, and currants) with which the settlers were already familiar, very few native plants were used as food.

My search of ethnobotanical reports and historical accounts has provided information on uses of edible prairie plants for these 17 tribes (see Map 2): Comanche, Apache, Arapaho, Cheyenne, Kiowa, Osage, Pawnee, Omaha, Ponca, Sioux (Dakota and Lakota), Crow, Arikara, Hidatsa, Mandan, Blackfoot, Assiniboin, and Plains Cree. There were an additional 14 tribes of Plains Indians for which no prairie plant-use information was found: Gros Ventre, Iowa, Kansa, Kitsai, Me'tis, Missouri, Oto, Plains Ojibwa, Quapaw, Sarcee, Tonkawa, and Wichita. Unfortunately, many facts about plant use were not recorded and are now

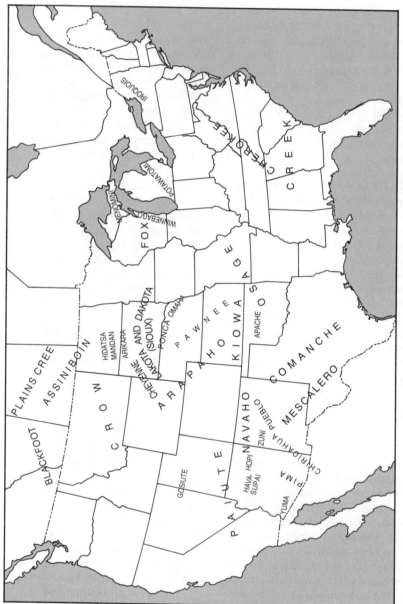

MAP 2. Indian tribes mentioned in this book that used prairie plants as food (locations c. A.D. 1700, before white settlement).

lost, because most Indian tribes of the region have been removed from their homelands and have given up their traditional food habits.

Other sources of information were reports about Indian tribes outside the Prairie Bioregion, including the Navaho, Havasupai, Mescalero and Chiricahua Apache, Pima, Yuma, and the Hopi, Zuni and other Pueblo tribes (Isleta, Acoma, Laguna, Cochiti, and Tewa)—all from the desert Southwest; the Gosiutes and Paiutes in the Great Basin; the Iroquois in the eastern woodlands; the Cherokee and Creek in the Southeast; the prehistoric Ozark Bluff-dwellers; and the Potawatomie, Fox (Meskaki), Menomini, and Winnebago, who lived along the Great Lakes. These tribes used plants whose geographical range extends into the prairie, so it is possible that Indians in the prairies used these plants in similar ways.

The Indians were dependent upon nature for most of their needs and had extensive knowledge of which plants were edible or useful. Because of this dependence, they believed strongly in a conservation ethic based on the sanctity of life. Melvin Gilmore studied the ethnobotany of the Omaha and Dakota (Sioux) and reported that the Indians "tell me they were taught by their parents and elders. . . . Do not needlessly destroy the flowers on the prairies

or in the woods. If the flowers are plucked there will be no flower babies (seeds); and if there be no flower babies then in time there will be no people of the flower nations. And if the flower nations die out of the world, then the earth will be sad. All the flower nations, and all the different nations of living things have their own proper place in the world, and the world would be incomplete and imperfect without them" (1977, pp. 97–98).

Although our culture is not directly dependent upon nature for its livelihood, we are beginning to see the problems created by depleting our natural resources. Perhaps by learning more about the native plants that surround us and about their use and history, we can begin to develop our own conservation ethic, which will bring us into harmony with our environment.

The greater part of the text comprises 49 plant chapters, each devoted to a particular species with significant food use. The plant chapters are followed by shorter descriptions of 11 grasses and 22 other edible plants with minor food uses. The discussions are arranged alphabetically by scientific plant name (common names, which might be more familiar to readers, are not consistent and often confusing).

Common names are provided to aid in recognition and because

they can give further information about the plants. These names are taken from the Great Plains Flora Association, 1986; Stevens, 1961; Steyermark, 1981 (1963); Nebraska Statewide Arboretum; Bailey and Bailey, 1976; and other sources cited. Indian names and their translations can give clues to the use, or the tribe's perception, of the plant. However, some tribal names can be translated only to English common names, adding nothing to our knowledge. Others are meaningless because the context in which they originated is not clear.

Each plant species described is listed under its scientific name as given in the *Flora of the Great Plains* (Great Plains Flora Assn., 1986). Derivation and meaning for these names come from Fernald, 1950; Stevens, 1961; Bare, 1979; and Bailey, 1962. The habitat of each plant is provided to aid in identification and to provide information on requirements for cultivation. This information is taken from the Great Plains Flora Assn., 1986; Stevens, 1961; Steyermark, 1963; and Bare, 1979.

Anyone who plans to search out and consume edible plants should exercise extreme caution. Correct identification of plants is necessary to avoid similar species or parts that may be unpalatable or poisonous. The best way to learn about plants is with the assistance of a well-informed friend or teacher and through the careful study of authoritative guides to plant identification. In the consumption of wild plants, abide by these five rules: (1) Know how to identify plants correctly. If there's any doubt, don't eat it. (2) Don't eat anything that doesn't taste good. (3) Eat new foods only in small amounts in case you have an allergic reaction. Some may be sensitive to the chemicals contained in and on various plants while others are unaffected. (4) Don't disguise the flavor when cooking, but don't hesitate to enhance flavors with salt, butter, sauces, sweeteners, or other seasonings. (5) Don't harvest plants from along roadsides or other areas that may have been treated with herbicides or affected by toxic emissions.

Differing opinions about the palatability of wild foods are common. It should be remembered that sense of taste varies greatly among individuals, that ripe fruit always tastes better than green, that flavor can vary from one plant to another, and that skillful preparation adds greatly to the enjoyment of any food. Nutritional information has been provided from the few studies conducted on edible prairie plants. More research on this subject would certainly highlight their value.

Many prairie plants are uncommon and some are rare. Wild plants should not be removed from their native habitats unless they are very abundant. This book notes which plants should be left undisturbed, to be appreciated as wildflowers. Fortunately, many edible prairie species can be propagated easily and grown in gardens for use as food or ornamentals without hurting the parent plants. Some general references on this subject include Bailey and Bailey, 1976; Steyermark, 1963; Stevens, 1961; and Salac et al., 1978.

Maps are provided to give the approximate geographical distribution of these plants. They were drawn from these sources: Great Plains Flora Association, 1977; Great Plains Flora Association, 1986; Steyermark, 1963; Elias and Dykeman, 1982; and Gleason and Cronquist, 1963. Some plants will occasionally be found outside the ranges in these maps; others will be hard to find within portions of their ranges.

This book is written for those who want to know more about prairie plants and the native food sources of the region. Whether your interest is stimulated by a springtime walk in a species-rich prairie preserve, or by driving down a country road lined with a dazzling array of prairie wildflowers, or by an interest in Plains Indians, I hope this book will prove both interesting and informative.

*Edible Wild Plants of
the Prairie*

Allium canadense
Wild Onion

C. Kuhn
© 86

Wild onion, wild garlic, prairie onion, meadow garlic, and Canada onion.

INDIAN NAMES

The Cheyenne call wild onions "kha-a'-mot-ot-ke-'wat" (skunk testes; probably a polite translation for something more derogatory) and "kha-ohk-tsi-me-is'-tse-hi" (skunk, it smells) (Grinnell, 1962, p. 171). The Lakota (Sioux) call wild onions "psin"; they specifically call *Allium drummondii* Regel. "ps'in s'ica'mna" (bad-smelling onion) (Rogers, 1980b, p. 25). The Osage name is "mon-zonxe" (earth, to bury) (Munson, 1981, p. 231). The Tewa, a Pueblo tribe in New Mexico, use the name "akonsi" (prairie onion or little prairie onion) (Robbins et al., 1916, p. 53). No translations were given for the following names. The Omaha and Ponca call wild onions "manzhonka-mantanaha"; the Winnebago, "shinhop"; and the Pawnee, "osidiwa" (Gilmore, 1977, p. 19). One name that the Blackfoot use is "pis-sats'e-mi-kim" (Johnston, 1970, p. 308).

SCIENTIFIC NAME

Al'lium canaden'se L. is a member of the Lilaceae (Lily Family). *Allium* is the ancient name for garlic. It is possibly derived from the Celtic "all," which means "pungent." The species name *canadense* means "of Canada."

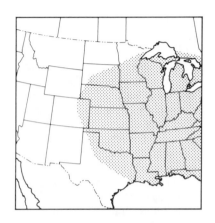

DESCRIPTION

Perennial herbs 2–9 dm (8–35 in) tall, growing from egg-shaped bulbs with a strong onion odor. Leaves 2 or more, basal, linear, 1–3 mm (1/16–1/8 in) wide, shorter than flowering stems. Flowers small, in round clusters at tops of erect stems, from Mar to Jul; 6 separate segments, 4–8 mm (3/16–5/16 in) long, white to pink or lilac, sometimes fragrant; flowers often replaced by bulblets. Fruits dry, small, round to oval, opening to release black seeds.

HABITAT

Prairies, roadsides, and open woods.

PARTS USED

Bulbs (late fall or early spring)—raw, cooked, or pickled; young leaves (spring, summer, or fall for *A. stellatum* Ker.)—either raw or cooked.

Wild onions are probably the best-known wild food. Several similar species of wild onions were valuable foods for the Indians, especially early in the spring, when they were used as greens and as a flavoring for other foods. The Cheyenne boiled onions with meat to enhance flavor, especially when there was no salt (Grinnell, 1962, p. 171).

Melvin Gilmore (1977, p. 19) reported that "all species of wild onion found within their habitat were used for food by the Nebraska tribes," including the Dakota, Omaha, Ponca, Winnebego, and Pawnee. The Blackfoot (in Montana) reportedly gather them in May or June and boil them with meat or preserve them for later use (Johnston, 1970, p. 308). Wild onions will keep a long time, because their skins dry and preserve the flesh inside. Buffalobird Woman reported that Hidatsa children dug and ate the raw bulbs in the spring. She compared this plant with the introduced onion and said that wild onions were never used for cooking (Nickel, 1974, p. 58).

The Comanche described two types of wild onions, a large sweet one, which they prepared by braiding the plants together and roasting them over a fire, and a smaller one with red flowers, which was considered less desirable (Carlson and Jones, 1939, p. 24). I believe that this large, sweet onion may

be Fraser's onion, *A. perdulce* S.V. Fraser. The Tewa, a Pueblo tribe in New Mexico, called the wild onion "prairie onion" or "little prairie onion" to distinguish it from the cultivated onion, which has become much more popular than the native species since its introduction (Robbins et al., 1916, p. 53).

The Menomini Indians of the Great Lakes region also thought that wild onions were skunklike. They called the wild leek, *A. tricoccum* Soland., "pikwu'tc sikaku'shia" (the skunk). "The word 'shika'ko' or 'skunk place' is the origin of the word Chicago, which in aboriginal times was the locality of an abundance of these wild leeks" (Fernald et al., 1958, p. 129). These abundant wild onions provided the primary source of food in this area for Father Jacques Marquette, the Jesuit missionary and explorer who traveled from Green Bay to Chicago in 1674 (Hedrick, 1919, p. 32).

Wild onions do not preserve well, but their archaeological remains have been found in rock shelters, which allow more vegetal materials to withstand degradation by the elements. Rock shelters are uncommon in the Prairie Bioregion, but remains found on the periphery of this area or in adjacent bioregions are significant and suggest that onions were used in the Prairie Bioregion as well. The outside husks of onions were found covering the floor of a rock shelter in central Wyoming that had a Middle Archaic occupation (approximately 3000 B.C.) (Morris et al., 1981, p. 215). Wild onions were found among the remains of the Ozark Bluff-dwellers (Gilmore, 1931b, p. 95). Also, they were found in 6,000-year-old human coprolites at Hind's Cave in the Lower Pecos area of southwest Texas. The remaining evidence of ingestion of the bulbs "ranged from minute fragments of the reticulate fibrous outer sheath or the veined papery inner tissues to more or less complete bulbs. The preservation in many instances is remarkable, resulting from the fact that the bulbs were frequently swallowed essentially unchewed" (Williams-Dean, 1978, p. 191).

Numerous explorers and early travelers made mention of the wild onions that grew in the prairies. Captain Meriwether Lewis, while traveling in late July 1805 up the Missouri River in an area of present-day western Montana where there was "little timber to obstruct [the] view," reported the following about wild onions:

I passed through a large Island which I found a beautiful level and fertile plain about 10 feet above the surface of the water and never overflown. on this Island I met with great quantities of smal onion about the size of a musquit ball and some even larger; they were white crisp and well flavored. I gathered about half a bushel of them before the canoes arrived. I halted the party
for breakfast and the men also geathered considerable quantities of those onions. it's seed had just arrived to maturirity and I gathered a good quantity of it. this appears to be a valuable plant inasmuch as it produces a large quantity to the squar foot and bears with ease the rigor of this climate, and withall I think it as pleasantly flavored as any species of that root I ever tasted. I called this beautifull and fertile island after this plant Onion Island (Thwaites, 1904, 2: 259).

Dr. E. James, botanist for the Long expedition, reported on June 7, 1820, as he left the Missouri River and entered the Platte River valley: "A species of onion, with a root about as large as an ounce ball, and bearing a conspicuous umbel of purple flowers, is very abundant about the streams, and furnished a valuable addition to our bill of fare" (McKelvey, 1955, p. 213). Later during the Long expedition (Thwaites, 1905, 14: 282–283) and also on the expedition of Prince Maximilian of the German kingdom of Wied, which explored the Missouri River region from 1832 to 1834 (Thwaites, 1906, 24: 82), the wild onion was eaten to cure a sickness that was thought to be scurvy, caused by a deficiency of vitamin C. Prince Maximilian and the others of his party who were ill ate the bulbs and leaves in the early spring, "cut up small like spinach," and quickly regained their health (ibid.).

It has been suggested that wild onions may have been a nutritious complement to the buffalo-meat diet of Indians in the area that is now northeastern Colorado (Morris et al., 1981, pp. 213–220). Onions contain large amounts of some important micronutrients, more vitamin C than an equal weight of oranges, and more than twice as much vitamin A as an equal weight of spinach (Zennie and Ogzewalla, 1977, pp. 77, 78). They also contain a significant quantity of starch, but it is mostly inulin, which is not easily digested by humans (Turner, 1981, p. 2349). The indigestibility of inulin (which occurs also in Jerusalem artichokes and wild hyacinth) is probably one reason why onions were not liked by some Indians and early explorers. Also, the smell was often considered to be rank.

Dr. V. Havard (1895, p. 113) of the United States Army reported: "The bulbs of all the species of *Allium*, or Garlic, are more or less edible and nutritious in spite of the strong-scented volatile oil they contain; many references are made to the 'Wild Leekes' and 'Wild Onions' by the first explorers who were sometimes compelled to follow the example of the Indians and eat them to sustain life; however, it was their abundance all over the land which gave them value rather than their quality."

On my walk from Kansas City to the Rocky Mountains, I learned from personal experience that wild onions (*A. drummondii*) may be hard to digest. While crossing a section of prairie that had not been "plowed out" in Cheyenne County in eastern Colorado, where recent corporate investment has tremendously changed the surface of the land, I stopped to eat some wild onions with a sandwich. Perhaps my stomach had already been turned a little by the disrespect the investors had shown for the fragile, wind-blown soils of the surrounding area. However, I soon felt worse—the result of heartburn from the 12 small, raw wild onion bulbs I had eaten.

One should always make certain that the onion plant being picked has the characteristic onion smell, to avoid confusion with the poisonous death camas, *Zygadenus nuttallii* Gray, which looks much like a wild onion but is odorless. It is described under wild hyacinth, another early spring plant that resembles the onion. These cautions are not intended to discourage people from eating properly identified wild onions. With their distinctive onion flavor, a few wild onions, diced fine, are especially appealing raw in a salad, sautéed with other foods, or added to soups and stews during the last 10 minutes of cooking time (if wild onions are cooked longer than that, they become slightly bitter). I also enjoy a few of their early greens in a salad.

16

Because wild onions are edible and most have attractive flowers, they are recommended for planting in an edible wildflower garden or some out-of-the-way place. They can easily be transplanted or they can be started from seed in a greenhouse early in the spring, after the seeds have been stratified.

The wild sweet onion or Fraser's onion is my favorite; I heartily recommend it for cultivation. It grows in sandy areas (where it is easy to harvest) of the central part of the Prairie Bioregion. It has a mild onion taste and is sweet and starchy, with a texture that reminds me of water chestnuts or Jerusalem artichokes. If we could grow enough of this onion, or if plant breeders could significantly increase the size of its bulbs, I believe that it could become an interesting new food source.

Amaranthus graecizans
Prostrate Pigweed

C. Kuhn
© 86

Prostrate pigweed, prostrate amaranth, spreading pigweed; other
species—pigweed, amaranth, care-
less weed, keerless weed, redroot,
tumbleweed, and quelite.

INDIAN NAMES

The Lakota call the prostrate pigweed "wahpe'-maka'-yata'pi-
iyec'eca" (leaves spread out on
ground); they call the pigweed,
Amaranthus retroflexus L.,
"yus'pu'-la-ota" (pulling off many
things with the hand), which may
refer to pulling out the sharp-
pointed sepals and bracts of pigweed from the skin after handling
the plant (Rogers, 1980a, p. 43).
The Zuni name for *A. graecizans*,
is "ku'shutsi" (many seeds), probably in reference to the seeds
being gathered (Stevenson, 1915,
p. 65).

SCIENTIFIC NAME

Amaran'thus grae'cizans L. and
A. retroflex'us L. are members of
the Amaranthaceae (Pigweed Family). *Amaranthus* means unfading
because the dry calyx and bracts
do not wither. The species name
graecizans means "stimulating";
retroflexus means "reflexed."

DESCRIPTION

Annual herbs with prostrate stems
1–6 dm (4–24 in) long, forming
mats. Leaves alternate, oval to elliptic or spoon-shaped, 8–40 mm
(⁵⁄₁₆–1½ in) long, pale green,

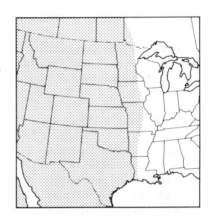

smooth, usually white-margined.
Flowers small, in dense clusters
among leaves, from Jun to Oct;
male and female separate, sepals
4, oblong, green with white margins, no petals. Fruits dry, roundish, 2.5–3.2 mm (± ⅛ in) long,
sometimes reddish, each containing 1 seed, lens-shaped, black, and
shiny.

A. retroflexus differs in being
upright (to 3 m or 10 ft tall) with
stout stems and having narrower,
longer leaves and flowers clustered densely on branches at tops
of plants.

HABITAT

Dry soil of prairies, fields, roadsides, and waste ground.

PARTS USED

Young plants (spring best)—
cooked; seeds (fall)—eaten raw,
cooked as cereal or mush, or used
as flour.

There are many species of pigweed that have been used for food. The prostrate pigweed is a common native species that will serve as a specific example. Pigweeds were more extensively used and reported as a food source (both for greens and seed) by the Indians in the Southwest and Great Basin regions than in the prairies because the drier conditions there made other food sources less available. During times of drought, this plant was probably more important for food across the Prairie Bioregion. Amaranth seeds have been found in archaeological remains throughout the region; it is believed that in some cases amaranth may have been cultivated.

J. W. Blankenship, in "Native Economic Plants of Montana" (1905, p. 6), reports that the prostrate pigweed was used as a pot herb and that seeds of this and other species were formerly used as food by Indians. The prostrate pigweed and the pigweed, *A. retroflexus*, were commonly used as food by the Navaho, Tewa, and other Pueblo Indians in the Southwest. They were boiled and eaten like spinach (Castetter, 1935, p. 25; Robbins et al., 1916, p. 53). The Apache also used these species of amaranths for greens (Opler, 1936, p. 46) and ate the seeds of several species (Reagan, 1929, p. 155).

Few stories concerning the food mythology and beliefs of the Indi-

ans of the Prairie Bioregion were recorded. The occurrence of amaranth in the tribal mythology of the Zuni shows its importance as a food plant and indicates its long-term use. The Zuni believed that the seeds of the prostrate pigweed were brought from the "undermost world" of the rain priests and were scattered by them over the earth (Stevenson, 1915, p. 65). The Zuni used these seeds as a food: "Originally the seeds were eaten raw, but the Zuni say that after they became possessed of corn, these seeds were ground with black corn meal, mixed with water, and the mixture was made into balls, or pats, and steamed, as are those eaten at the present time. A network of slender sticks or slats is fitted snugly inside the pot in the center, and the meal cakes or balls are placed thereon. The pot contains sufficient water to steam them" (ibid.).

Carolyn Niethammer, in *American Indian Food and Lore* (1974, p. 118), tells how a Havasupai woman demonstrated the harvesting of pigweed seeds in the traditional manner. She gathered the seeds by stripping the dried flower spike into her palm. Then she blew hard on the pile of fluff in her hand. The dried material flew off, leaving a small pile of black seeds.

The Yuman tribes that lived along the lower Colorado River in what is now southwest Arizona reportedly tied seed heads of pigweed together before they ripened, so they would be protected and no seeds would be lost (ibid., p. 119). The seeds of pigweed were harvested after the Yuman tribes had finished gathering their cultivated crops. Seed heads were broken off and taken home, where they were further dried. Threshing was done by beating the spikes with sticks on a hardened floor. The seeds were winnowed by tossing the crushed spikes in a basket, so that the wind could blow the chaff away. The seeds were then parched with coals in a shallow pottery dish, which caused the black seed coats to pop open, revealing the white inside (ibid.).

Prostrate pigweed and other species are "weedy." They aggressively take advantage of exposed soil in disturbed habitats, whether it be a prairie-dog town, a buffalo wallow, a stream bank, a roadside, or an agricultural field. Richard Ford, an ethnobotanist, believes that as the amount of agricultural land was expanded by Indian agriculturalists, the pigweed also expanded its range (1981b, p. 2186). Pigweed is not mentioned in the journals of Lewis and Clark or of other early explorers. This is probably because it is an inconspicuous plant and also because much less disturbed habitat existed at those times. This also explains why it is not mentioned in the historical materials on the Indians of the Prairie Bioregion. Nowadays, the pigweed is a major weed species in cultivated fields.

The leaves of pigweed are very perishable and do not preserve well. The seeds are very small, less than $\frac{1}{16}$ inch in diameter, and have previously been overlooked. Now, the seeds of pigweed are commonly found among other preserved plant materials at archaeological sites, which indicates their former importance as a food source. The oldest amaranth seeds found, located at a San Jose site (near Grants, New Mexico), were estimated to be 4,500 to 6,500 years old (Agogino and Feinhandler, 1957, p. 155).

Al Johnson, an archaeologist, believes that seeds of pigweed were gathered in the prairies by the Kansas City Hopewell during their occupation of Trowbridge Village from 200 to 500 A.D. (1981, pp. 71, 72). Pigweed seeds occur in such great quantity and with such frequency at the Mitchell Site (near present-day Mitchell, South Dakota) that there can be no doubt that they were collected intensively. This site was occupied from about 1250 to 1450 A.D. (Benn, 1974, p. 229). Seeds were also found at the Two Deer Site, near present-day El Dorado, Kansas, which was occupied from 800 to 1000 A.D. (Adair, 1984, p. 70).

CULTIVATION

Amaranth was a major cultivated plant of the Aztecs and was grown by some Indians of the Southwest. Montezuma II, the first Aztec emperor (1502–1520), received 200,000 bushels of cultivated

21

amaranth seed in annual tribute (Vietmeyer, 1981, p. 709). Melvin Gilmore (1931b, p. 97) suggests that the Ozark Bluff-dwellers (who lived adjacent to the Prairie Bioregion in northern Arkansas and southern Missouri) may have cultivated amaranth. Sheaves of the seed heads, which appeared to have been stored for future use as a seed source, were uncovered in their archaeological remains. It is also possible that these heads had been brought into their bluff dwellings to dry before threshing.

Edward Palmer reported in "Plants Used by the Indians of the United States" (1878, p. 603) that *A. retroflexus* and another species were "regularly cultivated by the Pah-Utes and are also found abundant in the wild state on river bottoms. The plants are very prolific in seeds, which are very nutritious and of an agreeable taste. Bread or mush made of the meal is very good and not to be despised."

Amaranth is again being grown as an agricultural crop for use in breads and cereals and is being tested for more extensive cultivation in the Midwest. It is a photosynthetically efficient plant that produces high yields of both greens and seeds. Grain yields of a domesticated species, *A. hypo-*

chondriacus, are greater than those of soybeans and only slightly smaller than those of wheat (Cole, 1979, pp. 193, 289). Amaranth greens have a very pleasant taste when young and tender and contain a significant amount of protein. In addition, the calcium, phosphorus, iron, potassium, vitamin A, and vitamin C contents are significant (Watt and Merrill, 1963, p. 6).

In a study of vegetable foods, including amaranth, extractable leaf protein was found to peak five to six weeks after planting (Oke, 1973, as cited in Bye, 1981, p. 116). This is the usual time of traditional Indian harvest for these greens, when they are several inches tall, but still tender.

Data from cornfields cultivated by the Tarahumara in Mexico show that the "weed crop" of *A. retroflexus* will yield 100 grams (3.5 ounces) of greens from plots varying from one to four square meters; in approximately one week there will be sufficient regrowth so that they may be harvested again (Bye, 1981, p. 114).

Amaranth is a food plant of the past that may become important in the future. Its seeds can be planted as an annual crop in the spring after the soil has warmed.

Ambrosia trifida
Giant Ragweed

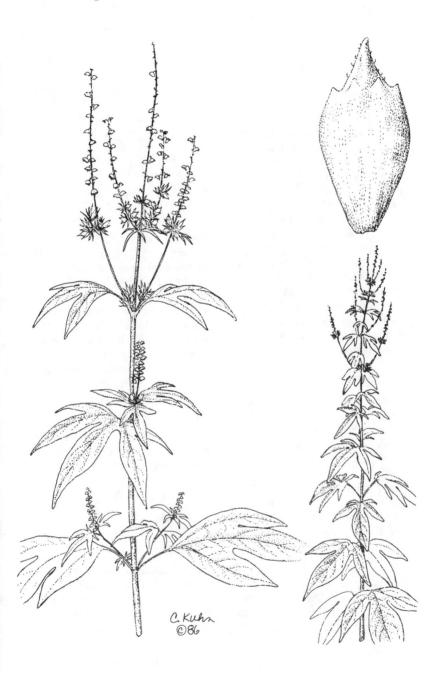

C. Kuhn
©86

COMMON NAMES

Giant ragweed, great ragweed, horse weed, and buffalo weed.

INDIAN NAMES

The Lakota call it "canhlo'gan was'te'mna" (bulky weed); they also call it "yamnu'mnuga iye'ceca" (it is like making noise crunching with teeth), in reference to the hard seeds that were used medicinally (Rogers, 1980a, p. 49). The Kiowa name (not given) literally translates as "bloody weed" (Vestal and Schultes, 1939, p. 55). The Kiowa children reportedly took great delight in breaking off a branch of this plant and watching the "bloody" liquid exude from the stems. Formerly, this plant was held in fear by the Kiowa people (ibid.).

SCIENTIFIC NAME

Ambro'sia tri'fida L. is a member of the Asteraceae (Sunflower Family). *Ambrosia* comes from the Greek "ambrotos," which means "immortal," in reference to several kinds of plants that were believed to have special properties. The species name *trifida* means "divided into three parts" and refers to the leaves.

DESCRIPTION

Annual herbs, stems ridged, 1–3 m (3–10 ft) tall. Leaves opposite, oval to roundish, 1–2 dm (4–8 in) long, most divided into 3–5 lobes,

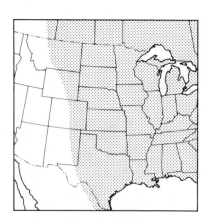

surfaces hairy, margins toothed. Flowers small, male in elongated groups at ends of branches, female in clusters among leaves, from Jul to Sep, greenish, nodding, yellow pollen conspicuous on male flowers. Fruits dry, egg-shaped, 6–12 mm (¼–½ in) long, each with a pointed beak and several knobby projections, ripening in Sep and Oct.

HABITAT

Fields, roadsides, disturbed places, and along railroads (often found in rich alluvial soils).

PARTS USED

Seeds (fall).

FOOD USE

Giant ragweed is the bane of allergy sufferers who are affected by hay fever in the fall. Because of its bitter leaves and tall, rank stature, it would seem to be an unlikely

24

food plant. However, its seeds were used by prehistoric Indians; apparently, the plant was even cultivated. It is not recommended because little is known about how to process and eat the seeds, but it is included here because of its history.

CULTIVATION

Melvin Gilmore (1931b, p. 85) was the first to mention giant ragweed as a food source of prehistoric Indians. In his analysis of the remains of the Ozark Bluff-dwellers, he reported: "There is evidence that the ancient Ozark Bluff-Dwellers also had certain other species of plants not cultivated at the present time. The ground for this statement lies in the fact that supplies of the seed of these other species of plants were carefully put away together with the selected seed of corn, beans, sunflowers, squashes and pumpkins." These ancient crops included a lamb's quarters (*Chenopodium*), a pigweed (*Amaranthus*), marsh elder (*Iva annua*), maygrass (*Phalaris caroliniana*), and giant ragweed (*Ambrosia trifida*) (ibid.).

Gilmore (ibid., p. 86) also described the seeds of cultivated giant ragweed as being larger than any found growing wild (sometimes four or five times larger), lighter colored, and in great abundance. A review of the archaeological significance of giant ragweed by Willard Payne and Volney Jones (1962, pp. 150, 162) showed that its seeds are quite variable in size; that the Ozark Bluff-dwellers' samples fall within that range of variability; that Gilmore's reaction to the size, color, and abundance was overstated; and that storage of seeds in a cave or bluff shelter for next year's crop is questionable because of the dampness of this environment. They concluded that giant ragweed was gathered from wild stands, rather than actually having been cultivated, and that it may have been primarily used as medicine, rather than as a source of food. Waldo Wedel (1955, p. 145) reported that giant ragweed was found in the archaeological remains from the vicinity of Mobridge, South Dakota, and also suggested that it may have been used for medicine.

Most recently, David and Nancy Asch have resurrected giant ragweed as an economic, cultivated species, grown during the Archaic Period in west-central Illinois and abandoned prior to Woodland times (about 1500 B.C.) (1982, p. 11). Their case for the economic utilization of giant ragweed rests on the facts that almost without exception, the adherent shell is missing from the carbonized seeds (suggesting that the seeds had already been processed) and that it is abundant in several Archaic seed samples. In particular, giant ragweed seeds ranked second to the cultivated marsh elder (*Iva annua*) in abundance and constituted 20 percent of all identifiable seeds for Horizon 6 at the Koster site (along the Illinois River in southwest Illinois) (ibid.).

The irritating pollen was probably a significant factor in the abandonment of ragweed as a cultivated crop. Indians who were allergic to this pollen and were living near a field of ragweed must have suffered severe reactions.

Amelanchier alnifolia
Serviceberry

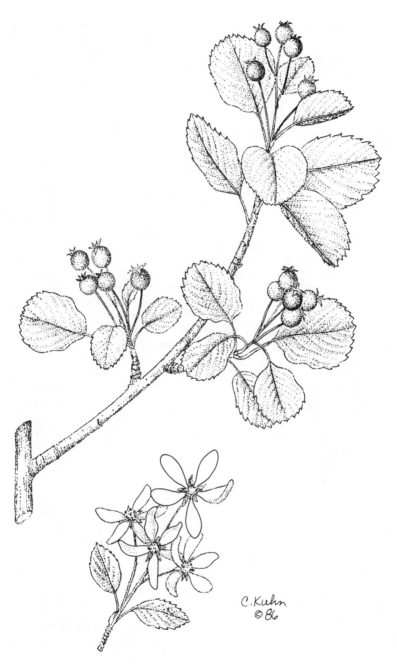

C. Kuhn
© 86

Serviceberry, saskatoon, saskatoon serviceberry, saskatoon berry, common serviceberry, western serviceberry, sarvisberry, alderleaf sarvisberry, juneberry, western juneberry, pigeonberry, shad, shad-bush, shadberries, shadblow, shad-blow serviceberry, comier, sugar pear, mountain pear, and Indian pear.

INDIAN NAMES

The Blackfoot name is "ok-kun-okin" (berry) (McClintock, 1909, p. 277). The Cheyenne name is "he-tan-i-mins" (male berry), which suggests that it has strong qualities (Grinnell, 1962, p. 176). The Lakota called it "wipanzut-kan," which refers to a thing used to crack bones (Rogers, 1980b, p. 56). The Plains Cree call it (mu-saskwatonina) (Mandelbaum, 1940, p. 203) and the Assiniboin called it "we-pah-zoo-kah" (De-nig, 1930, p. 583). The Omaha and Ponca name is "zhon-huda" (gray wood); and the Winnebago name is "haz-shutsh" (red-fruit) (Gil-more, 1977, p. 35).

SCIENTIFIC NAME

Amelanch'ier alnifol'ia Nutt. is a member of the Rosaceae (Rose Family). *Amelanchier* is the French Savoy word for the medlar, *Mespilus germanica*, which has similar fruits. The species name *alnifolia* means "alder-like leaves."

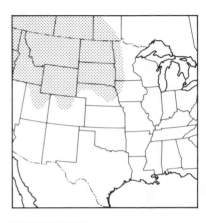

DESCRIPTION

Shrubs 1–3 m (3–10 ft) tall, young branches hairy. Leaves alternate, broadly elliptic to 4-angled, 2.5–5 cm (1–2 in) long, longitudinally folded at flowering, lower surfaces with yellowish hairs, margins toothed mostly above middle. Flowers in elongated groups of 3–20, from late Apr to May; petals 5, separate, egg-shaped to spoon-shaped, 6–8 mm (¼–5⁄16 in) long, white to pinkish. Fruits fleshy, round, 8–11 mm (5⁄16–7⁄16 in) in diam, dark purple, sweet, and juicy, ripening Jun through Aug.

HABITAT

Brushy hillsides, open woods, creek banks, usually in well-drained soil, but occasionally around bogs.

PARTS USED

Ripe fruit (summer—raw, cooked, or dried; leaves (spring, sum-mer)—dried for tea.

The serviceberry or saskatoon was a major food of native people in the northern prairies. It occurred widely and in varied habitats. It was probably the most important vegetable food of the Blackfoot. They used the berries in great quantities, fresh or dried, in soups and stews, and with meats (Johnston, 1970, p. 313). Because of the great importance of the serviceberry as a food source, the Blackfoot moved their midsummer camp to a location where it could easily be harvested. The women and children beat the bushes with sticks, causing the berries to fall on robes or blankets that had been spread out on the ground (Ewers, 1958, p. 86).

The most common use of the serviceberry by the Blackfoot was for pemmican, which was made from a mixture of animal fat, dried meat, and the crushed berries. For a sausagelike food, serviceberries were mashed along with an equal amount of fat and stuffed into an animal intestine, which was tied at both ends and boiled (Hellson, 1974, p. 100). For food reserves, serviceberies were prepared thus: "The berries were dried on a raised hide that had numerous holes punctured in it. The hide was kept four or five feet from the ground to discourage children and animals from eating the berries. When it was dry, the choice of the harvest was greased with backfat and stored in hide bags made of fetal deer or sheep. That way the berries would not spoil. Then they would be added to soups or made into sausages" (ibid.).

A favorite dessert at Blackfoot feasts was a soup made from buffalo fat and berries mixed with buffalo blood (Johnston, 1970, p. 313). Serviceberry leaves were also used. They were crushed and mixed with blood, then dried and used to make a rich broth in winter. George Bird Grinnell (1962, p. 176) reported that the serviceberry had a sacred significance among the Blackfoot. His observation is supported by the facts that serviceberries were a favorite snack reserved for men (Hellson, 1974, p. 100) and that a forked stick from the serviceberry bush was used in religious rituals (Johnston, 1970, p. 313). The many geographical names in Alberta and Montana containing various Blackfoot translations for serviceberry in them, e.g., Many Berries Creek, Many Berries Coulee, Cherry Coulee, and Many Cherry Bushes Valley, indicate the importance of these berries in their perception of the world (ibid.).

The Assiniboin started harvesting serviceberries at the end of the season when prairie turnip (*Psoralea esculenta*) roots were dug, and all who were able-bodied helped with the berry harvest. "Although men never helped with any household duties, it was no disgrace for them to pick serviceberries, as there was always danger of bears in the berry patches. So the men acted as escorts and sometimes helped pick the berries" (Kennedy, 1961, p. 82). The berries were spread out on hides, dried out in the sun, and packed away in berry bags. "The grandmothers usually tanned fawnskins for use as berry bags, the skins from speckle-backed fawn being the most popular. When cured, these skins were tanned with the hair left on, and all holes were sewed shut, except for one opening in the under side. The skin was filled and packed with cured berries and the opening sewed up. When completed, the bag resembled a stuffed fawn. It was then presented to a favorite grandchild" (ibid.).

The Assiniboin often rubbed fresh serviceberries into pounded dried prairie turnips (also called breadroot). This mixture was dried in the sun and stored away for winter food (Kennedy, 1961, p. 82). H. Y. Hind reported having eaten a pudding of this mixture among the Plains Cree in 1859, which he described as being "very palatable" (Wedel, 1978, p. 13).

Serviceberries were collected in late June or July by Arikara women and were sun dried for later use. "Each woman filled one or two 'calf skin bags' during a day's collecting. These were dried on the corn drying stage, and occasionally winnowed to eliminate leaves and stem parts. Dried berries were bagged and either 'cached' or suspended within the lodge. Dried berries were added to

pounded, dried breadroot and mixed in the mortar with some bone grease or broth. Both blueish and white fruited forms were collected and used in the same manner" (Nickel, 1974, p. 58).

The Arikara also bought serviceberries from other tribes, paying two measures of shelled corn for one measure of dried serviceberries. Chokecherries were the same price. Serviceberries "are easier to prepare for drying than chokecherries, but harder to gather. The chokecherries are easy to gather, but the process of pounding to a pulp and drying is laborious, hence they were equalized in price" (Gilmore, 1926, pp. 15–16).

The Cheyenne gathered serviceberries in quantity to dry and store for winter use in stews. Serviceberries were also an important article in feasts. A tea made from the leaves was used by the Cheyenne as a beverage and a medicine. It was red in color and reportedly had a flat taste like green tea (Grinnell, 1962, p. 176).

The Omaha, Ponca, Winnebago, and Sioux also ate serviceberries (Gilmore, 1977, p. 35). The Lakota call the equivalent of our month of June "Wi'pazuka-waste'-wi," which means "good Juneberry (serviceberry) moon" (Rogers, 1980a, p. 90).

Serviceberries have been used in the northern prairies since antiquity; their pits have been found at Walth Bay, a prehistoric site near Mobridge, South Dakota (Nickel, 1974, p. 38). Explorers, trappers, and traders commonly used the serviceberry as food. John James Audubon, the famous painter of birds, made a trip up the Missouri River in 1843. On July 26, when his party camped about 30 miles downstream from Fort Union, which was located at the junction of the Missouri and Yellowstone rivers (in present-day Montana), he reported: "We found plenty of water, and a delightful spot, where we were all soon at work unsaddling our horses and mules, bringing wood for fires, and picking service-berries, which we found in great quantities and very good" (McFarling, 1955, p. 190).

While in the Rocky Mountains, Sergeant Patrick Gass of the Lewis and Clark expedition reported on August 22, 1805, another service-berry mixture that the Indians ate: "The people of these three lodges have gathered a quantity of sun-flower seed, and also of lambs-quarter, which they pound and mix with service-berries, and make of the composition a kind of bread, which appears capable of sustaining life for some time. On this bread and the fish they take out of the river, . . . these people chiefly subsist" (Thwaites, 1905, 3: 16n).

David Thompson, another early explorer, wrote: "On the Great Plains there is a shrub bearing a very sweet berry of a dark blue color, much sought after, great quantities are dried by the Na-tives; in this state, these berries

are as sweet as the best currants and as much as possible mixed to make Pemmecan" (Tyrrell, 1916, in Johnston, 1970, p. 313).

Early trappers and traders pur-chased pemmican for winter pro-visions from the Indians at trading posts owned by the Hudson Bay Company and the North West Company. In 1811, at the North West Company Trading Head-quarters along the Red River (in present-day southeast Manitoba), there were several hundred sacks of pemmican laid up in the store-house. The following report de-scribes how this pemmican was made:

Pemmican, *of which I have al-ready spoken several times, is the Indian name for the dried and pounded meat which the natives sell to the traders. About fifty pounds of this meat is placed in a trough, and about an equal quan-tity of tallow is melted and poured over it; it is thoroughly mixed into one mass, and when cold, is put up in bags made of undressed buffalo hide, with the hair outside and sewed up tightly as possible. The meat thus im-pregnated with tallow, hardens, and will keep for years. It is eaten without any other preparation; but sometimes wild pears (ser-viceberries) or dried berries are added, which render the flavor more agreeable (Thwaites, 1904, 6: 380n).*

Nutritious serviceberries were not only used to add flavor to

31

pemmican and other foods, but were enjoyed by themselves. J. W. Blankenship reported in "Native Economic Plants of Montana" (1905, p. 6) that serviceberries "are found frequently in our markets during the season and are used for pies, much as the blueberries are used in the East; they are usually mixed with currants, gooseberries or rhubarb, to lend acidity to the combination. They are also put up spiced, are used for wine and made into jam with other fruits." Serviceberries contain more than three times the amount of iron and copper in the same weight of raisins (Rivera, 1949, in Turner, 1981, p. 2334).

CULTIVATION

The serviceberry is one of the most ornamental native woody plants when it is densely covered with snowy white flowers. It blooms in early spring, while other trees and shrubs are still leafless. It should be planted for its beauty as well as for the fruits, which benefit both man and wildlife. Serviceberry has been suggested as a possible new commercial fruit crop for Canada (Graham, 1977, in Turner, 1981, p. 2333). It is extremely winter hardy and is a good fruit for the northern prairies where the winters are cold and rainfall is low. There are numerous cultivars, including "Shannon" and "Indian," which are superior to the wild species in having larger fruits and being more productive. "Atlaglow" is excellent for ornamental purposes and its fruit is quite sweet. Serviceberries are propagated by seeds or by suckers.

Amorpha canescens
Leadplant

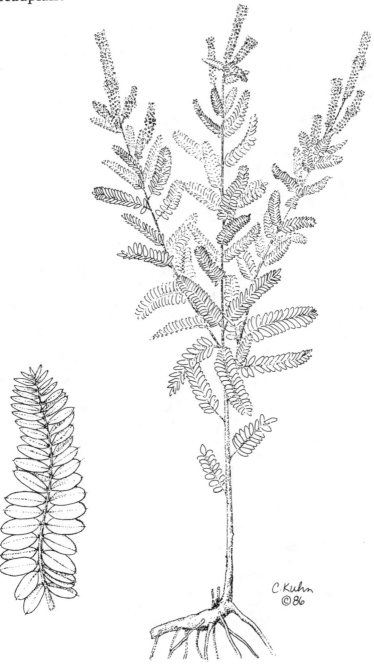

C. Kuhn
© 86

Leadplant (from the leaden color of the foliage), tea plant, and prairie shoestrings. (Many of the tough, woody roots of this plant are spread out parallel to and just below the soil surface. When settlers were plowing the prairies to plant crops, these roots would break with a characteristic pop, like a shoestring breaking, which gave rise to this common name) (Weaver and Fitzpatrick, 1934, p. 200).

INDIAN NAMES

The Omaha and Ponca call this plant "te-huntonhi" (Buffalo bellow plant) because it is the dominant prairie plant in flower during the rutting season of the buffalo (Gilmore, 1977, p. 41). The plant was considered to be female, while roundhead lespedeza (*Lespedeza capitata* Michx.) was the "male buffalo bellow plant" (ibid.). The Lakota name is "zitka'tacan'" (the bird's wood or the bird's tree) because birds perch on it in the prairie where there are no trees (Rogers, 1980b, p. 70).

SCIENTIFIC NAME

Amor'pha canes'cens Pursh is a member of the Fabaceae (Bean Family). *Amorpha* comes from the Greek "amorphos," which means deformed, referring to the incomplete flower structure (the wings and keel of the corolla are absent). The word *canescens* means "be-

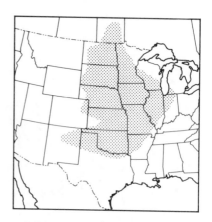

coming gray," referring to the thick covering of fine hairs that gives the plant a grayish appearance.

DESCRIPTION

Shrubs 3–8 dm (1–2½ ft) tall, stems 1 to several, often branched. Leaves alternate, pinnately compound, 13–20 pairs of leaflets plus 1, leaflets oblong to elliptic, usually hairy, with brownish, sharp tips. Flowers in elongated groups among upper leaves, forming dense clusters, from May to Aug; 1 petal, heart-shaped, curved, bright violet. Fruits dry, oblong, 3.5–4 mm (±³⁄₁₆ in) long, surface covered with whitish hairs and glands, opening lengthwise to release seeds, olive brown and smooth.

HABITAT

Prairies, plains, and rocky open woods.

34

Leaves (late spring and summer)—
dried and used for tea.

FOOD USE

Leadplant is an important legume
in the tallgrass and midgrass prairie ecosystems, but only a minor
food source. The Dakota sometimes used the leaves to make tea
(Gilmore, 1977). Tea made from
the dried leaves is yellow and
pleasant to taste.

Leadplant is most useful as an
indicator of diverse prairie, where
other edible plants can be found.
In most areas of the tallgrass prairie, the absence of leadplant indicates that the original prairie vegetation has been significantly
disturbed. Leadplant is much less
common even on grasslands than
it was before settlement, because
it is very palatable to livestock
and decreases under heavy grazing.

John Charles Frémont reported
on June 20, 1842, while in northeast Kansas: "Along our route, the
amorpha has been in very abundant but variable bloom: in some
places, bending beneath the
weight of purple clusters; in others, without a flower. It seems to
love best the sunny slopes, with a
dark soil and southern exposure"
(Jackson and Spence, 1970, p.
177). Two days later, near Sandy
Creek in southeast Nebraska, he
reported: "The country has become very sandy, and the plants
less varied and abundant, with the
exception of the amorpha, which
rivals the grass in quantity, though
not so forward as it has been
found to the eastward" (ibid., p.
178).

The abundance of leadplant was
also noticed by Lieutenant J. W.
Abert in 1846. While headed west
to Bent's Fort, he reported that
along the Pawnee Fork of the Arkansas River (in west-central Kansas) the tea or lead plant "is in
some places so abundant as to dis-
place almost every other herb"
(Malin, 1961, p. 117).

Leadplant is a perennial with a
woody base from which coarse annual stems arise. In places that
have not been mowed or heavily
grazed, the plants actually develop
into bushes 2½ to 4 feet tall, with
widely spreading stems. These
taller shrubs were probably more
common in presettlement times,
although fires and heavy local
grazing by buffalo would periodi-
cally shorten them. We can imagine that the prairie had a much
different appearance then. John
Weaver studied a large number of
upland prairies and documented
leadplant as being the most important plant, other than the
grasses. In a few limited areas,
leadplant was even more abundant
than the grasses, with up to 60
plants per square meter (Weaver
and Fitzpatrick, 1934, p. 195).

As with most perennial prairie
plants, there is a larger portion of
leadplant below ground than
above. Its extensive woody root
system extends to a depth of 6½

to 16½ feet. The total root system penetrates a substantial volume of soil, as numerous lateral roots often spread out 4 feet or more from the crown (ibid.). Leadplant, like other legumes, has bacterial nodules on its roots that can assimilate from the atmosphere the nitrogen so important for plant growth. These nodules play a significant role in the cycling of nitrogen in the prairie ecosystem. They occur along the entire length of the roots of leadplant (ibid., p. 199).

CULTIVATION

Leadplant is quite showy, with its silvery-gray foliage and dark purple flower spikes. Julian Steyermark (1981, p. 902) reports that it is sometimes grown as an ornamental plant and adapts itself well to sunny dry situations. It can be grown from seed or softwood stem cuttings. Leadplant has about 123,000 seeds per pound. When mixed with native grass seed, only about three leadplant seeds need to be planted per square foot to produce an adequate stand (Salac et al., 1978, pp. 4, 8). Leadplant is useful in prairie restorations and also can be grown as a teaplant in a wildflower garden.

Amphicarpaea bracteata
Hog Peanut

C. Kuhn
©86

COMMON NAMES

Hog peanut, wild peanut, pea vine, and Dakota peas.

INDIAN NAMES

The Dakota name is "maka-ta-omnicha" or "onmnicha" (ground beans); the Omaha and Ponca say "hinbthi-abe" (beans); the Pawnee name is "ati-kuraru" (ground beans); and the Winnebago call it "honink-boije" (Gilmore, 1977, p. 43).

SCIENTIFIC NAME

Amphicar'paea bractea'ta (L.) Fern. is a member of the Fabaceae (Bean Family). *Amphicarpaea* comes from Greek and means "with two kinds of fruit," referring to the fruits produced both above and below the ground. The species name *bracteata* means "small leaf," referring to the flowers being subtended by small leaves or bracts.

DESCRIPTION

Annual herbs growing from tap-roots, stems 3–20 dm (1–6½ ft) long, twining on shrubs or on the ground. Leaves alternate, divided into 3 leaflets, oval to 4-angled, 2–10 cm (¹⁄₁₆–³⁄₈ in) long, sometimes hairy. Flowers of two kinds, from Aug to Oct; some in elongated groups among leaves, 5 petals, upper 1 large and erect, lower 2 boat-shaped, and 2 wings at sides, lilac or white; others near base of stem or underground, lacking well-

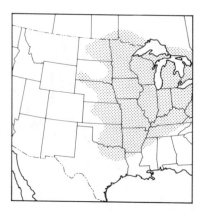

developed petals. Fruits above-ground dry, flattened, 1.5–4 cm (⁵⁄₈–1⁵⁄₈ in) long, sickle-shaped, seeds kidney-shaped and brown; underground fruits fleshy, 6–12 mm (¼–½ in) in diam, not open-ing and each containing 1 seed.

HABITAT

Woods and thickets.

PARTS USED

Underground seeds (late fall, early spring)—raw or boiled; above-ground seeds (fall)—boiled.

FOOD USE

This unusual vine is found in thickets and woods that extend into the prairies. Its subterranean fruits were used for food by many of the Indians of this region. In the *Social Life of the Omaha Tribe,* by Alice Fletcher and Fran-cis La Flesche (1911, p. 341), it is reported that the Omaha har-vested hog peanut "roots." (Actually, the hog peanut is an underground fruit produced after the flower blooms and buries itself like a peanut.) These "roots" were gathered "in the fall from the storehouses of the field mouse. This little animal gathers these roots in large quantities. The Indians kept the roots in skin bags during the winter. Before boiling, the outer skin was removed by rubbing the root between the palms of the hands. The flesh is whitish before cooking and reddish afterward; it is sweetish in taste and very nutritious."

Field mice gathered large quantities of hog peanuts (from a pint to several quarts) in places where they were plentiful (Gilmore, 1913a, 1977). Melvin Gilmore, an ethnobotanist, found a strong popular feeling of affection and respect for the bean mouse among all the tribes of the Missouri river region. The Omaha expressed the sentiment that "the bean mice are very industrious people; they even help human beings" (Gilmore, 1925, pp. 182–183).

The Ponca were related to the Omaha and spoke the same language. Both tribes raised beans in their gardens, but they also utilized the stores of wild beans that had been collected by rodents and stored by the animals in their burrows (Howard, 1965, p. 44).

The Pawnee formerly inhabited a large part of Nebraska, with villages on the Loup, Platte, and Republican rivers. In 1875, this tribe was removed to a reservation in Oklahoma. James R. Murie, a Pawnee, wrote to Melvin Gilmore in 1913 about a hog peanut specimen that Gilmore had sent him: "We call them *atikuraru*. . . . The Pawnees ate them. In winter time the women robbed rats' nests and got big piles of them. Nowadays when the old women see lima beans they say they look like *atikuraru* in Nebraska" (Gilmore, 1977, p. 44).

The hog peanut was used by the various bands of the Sioux. Philander Prescott stated in *Farming Among the Sioux Indians* (1849, p. 452) that the Santee Dakota in the Minnesota Territory used the hog peanut, which is found "in all parts of the valleys where the land is moist and rich. It is the size of a large white bean, with a rich and very pleasant flavor. When used in a stew, I have thought it superior to any garden vegetable I have ever tasted. The Indians are very fond of them, and pigeons get fat on them in the spring. . . . The beans on the ground are gathered by the Indians, who sometimes find a peck at once, gathered by mice for their winter store."

As recently as 1934, the Santee Dakota women of Prairie Island (Minnesota) still dug the wild bean or hog peanut. One of them, Grace Rouillard, described this work: "Women did not usually go directly to the . . . bean patches, but early in September they raided the storehouses of field mice on the offchance of finding what they sought. If no mouse store appeared, the women turned up the earth with a digging stick about three feet long, having a bowl like a spoon, made of ironwood; they plunged it in at a slant, with both hands. Men or women made this tool, burning a bowl to harden it. A woman dug up a big area to find enough vegetables for winter" (Landes, 1968, p. 203).

The bean mice were respected by the Dakota and the women would leave them a gift or payment when taking hog peanuts from their stores. Melvin Gilmore (1921, p. 607) reported that "in all accounts I have had from the people of the Dakota nation the women have always said that they never took away any beans from the voles without making some payment in kind. They said it would be wicked and unjust to take the beans from the animals and give nothing in return. So they said they always put back some corn, some suet, or some other food material in exchange for the beans they took out. In that way they said both they and the little animals obtained a variety in their food supply."

The Dakota have a popular moral tale that exemplifies their attitude toward the bean mouse:

A certain woman plundered the storehouse of some Hintunka people (bean mice). She robbed them of their entire food supply without giving them anything in return. The next night this woman heard a woman down in the woods crying and saying, "Oh, what will my poor children do now?" It was the voice of the Hintunka woman crying over her hungry children.

The same night the unjust woman who had done the wrong had a dream. In her dream Hunka, the spirit of kinship of all life, appeared to her and said: "You should not have taken the food from the Hintunka people. Take back the food to them, or some other in its place, or else your own children shall cry from hunger."

Next morning the woman told her husband of this vision, and he said "You would better do as Hunka tells you to do." But the woman was hardhearted and perverse, and would not make restitution for the wrong she had done.

A short time afterward a great prairie-fire came, driven by a strong wind, and swept over the place where the unjust woman and her family were camping. The fire consumed her tipi and everything it contained, and the people barely escaped with their lives. They had no food nor shelter; they wandered destitute on the prairie, and the children cried from hunger (Gilmore, 1925, pp. 183–184).

Women of the Dakota Nation also gathered the small bean seeds produced in quantity in the upper branches of the hog peanut plant. These seeds are about the size of lentils (Gilmore, 1977, p. 44).

Different cultures have different relationships to plants, animals, and food. The difference between white American culture and Indian culture is great and observations of Indian cultures by whites have often been racist. An example is this account by F. V. Hayden (1862, p. 370) of one way that "Dakota peas" (hog peanuts) were served: "Some of the dishes prepared by the Indians in the yet undeveloped condition of their culinary science are not enticing even to the eye of the hungry traveler, and are by no means adapted to the delicate stomachs or fastidious palates. . . . The prairie turnip boiled with the dried stomach of the buffalo, or the Dakota peas abstracted from a mouse's nest and cooked with dried beaver's tail or a fat dog, are dishes much admired and regarded fit to set before soldiers, chiefs, and distinguished visitors."

Several early travelers and explorers mention the use of the hog peanut by Indians, but almost all of them are vague and uncertain as to the nature and identity of the plant and the animal that harvests it. While among the Arikara, Captain William Clark of the Lewis and Clark expedition reported on October 11, 1804, that: "Those people gave us to eate bread made of Corn & Beans, also Corn & Beans boiled, a large Been

(of) which they rob the mice of the Prairie (who collect and discover it) which is rich & verry nurrishing." (Thwaites, 1904, 1:187).

Father de Smet was an indefatigable Christian missionary who traveled widely across the upper Missouri River region. He wrote in his journal on August 1, 1851, of the wild foods of the region and specifically about the hog peanut: "Peanuts are also a delicious and nourishing root, found commonly in low and alluvial lands. The above-named roots form a considerable portion of the sustenance of these Indians during winter. They seek them in places where the mice and other little animals, in particular the ground-squirrel, have piled them in heaps" (McFarling, 1955, p. 204).

The bean mouse was identified as the vole, *Microtus pennsylvanicus*. It is about 5½ inches long and weighs up to 2½ ounces (Gilmore, 1925, p. 181). This vole hollows out a place in the ground and covers its hoard with sticks, leaves, and earth. The Indians reported that it uses a leaf of the boxelder tree, or sometimes another kind of suitably shaped leaf, as a sled for gathering hog peanuts (Gilmore, 1921, pp. 607, 609).

CULTIVATION

The hog peanut was probably encouraged in semiwild stands, rather than actually being cultivated by the Indians. It may have been cultivated in the South as a vegetable, as reported in 1869 by F. P. Porcher in *Resources of the Southern Fields and Forests* (Hedrick, 1919, p. 46).

The hog peanut (also called the ground bean) was promoted by Melvin Gilmore as a potential crop for farmers. According to Addison E. Sheldon (1923, p. 79), the superintendent of the Nebraska State Historical Society, Gilmore stated:

When asked why the experiment stations of the agricultural schools have not tried the development of the ground bean the botanist replied: "Bah! That's too simple and practical a problem. They would rather spend time in trying to coax something to grow that is foreign to the climate than develop what nature already has acclimated."

The food value of the native ground bean is such that Mr. Gilmore expresses a strong criticism of the neglect of the white men to give it a place among cultivated plants. In its wild state it is ahead of our cultivated beans, he declares. If developed by the processes of seed selection and cultivation we have no idea of what a food product might have resulted. Belonging to the legume family it enriches the soil with nitrates.

The hog peanut is still being promoted as a potential "root" crop for gardens (Dore, 1970, pp. 7–11). Its seeds are nutritious. A sample of the underground fruit harvested in Michigan contained 22.1 percent starch and 24.5 percent protein (Yanovksy and Kingsbury, 1938, p. 657). It can be grown from either of the two types of seeds. Hog peanut does not need full sun but prefers rich, moist soil.

Androstephium caeruleum
Blue Funnel Lily

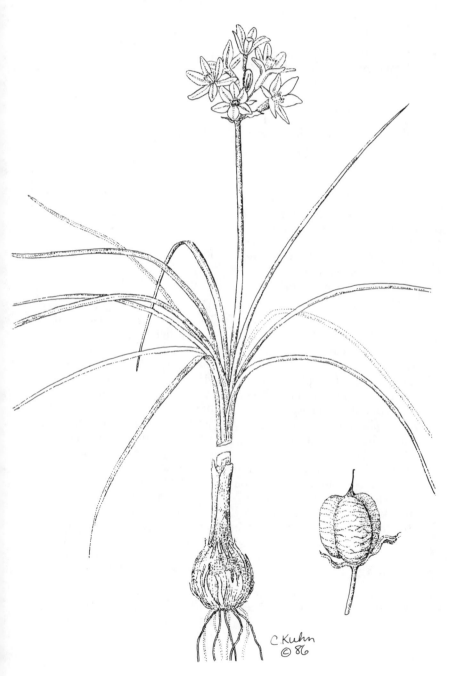

C Kuhn
© 86

Blue funnel lily and funnel lily.
(Funnel refers to the shape of the
flower.)

INDIAN NAMES

None were found in the sources
consulted.

SCIENTIFIC NAME

Androsteph'ium caerul'eum
(Scheele) Torr. is a member of the
Liliaceae (Lily Family). *Andro-
stephium* comes from Greek and
means "man wreath," referring to
the arrangement of the flowers.
The species name *caeruleum*
means "sky blue."

DESCRIPTION

Perennial herbs 1.5–3 dm (6–12
in) tall, growing from corms.
Leaves 5 or 6, basal, linear, 2–3
mm (¹⁄₁₆–⅛ in) wide, longer than
flowering stems. Flowers small, in
round clusters at tops of erect
stems, in Apr or May; each 1.8–3
cm (¾–1¼ in) long, 6 segments
united at base and separating into
oblong lobes, tube and lobes about
same length, pale blue to violet,
fragrant. Fruits dry, 3-angled, 12–
16 mm (½–⅝ in) long, opening to
release black seeds.

HABITAT

This species occurs only on prairies from central Kansas to Texas.

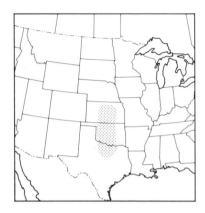

PARTS USED

Root (spring).

FOOD USE

This beautiful spring wildflower
has a limited range of distribution, is uncommon within that
range, and should not be harvested
from the wild as a food source.
Henry H. Rusby (1906b, p. 220)
gave the only account of the blue
funnel lily being used as a food
source. He reported that the corm,
which is composed of layers of
concentric circles, was eaten in
west Texas.

The blue funnel lily has attractive flowers, which are quite fragrant and smell like Concord
grape juice. The corm is unusual
and has been described by William Chase Stevens in *Kansas
Wild Flowers* (1961, p. 29):

*The first year there is a tiny
bulblet bearing a single leaf and a
few rootlets. Then follows in suc-
ceeding years the production of a*

fleshy turgid taproot, stored mostly with sap. The corm ultimately resides about 3 inches below the surface, being brought to that depth by the contraction of the turgid root, which is annually renewed until the corm arrives at its proper depth. Examining an entire plant at the time of its blooming in April one finds that the bulb consists of an upper and a lower segment, the latter producing a ring of roots around its equatorial circumference (not at the bottom, as portrayed in some manuals), the upper segment bearing the leaves and inflorescence. The following April the bottom segment will have disintegrated, the top segment will have become the bottom segment, bearing a new tip segment and a new turgid taproot, if the normal depth of the corm has not yet been reached. Thus the plant never grows old, no part of it ever living beyond its second year. The perennial nature of this plant is inherent in its growing point which each year, while producing a new corm segment, with leaves and inflorescence, is at the same time replacing every cell of its structure by new cells.

The blue funnel lily is among the prairie plants that renew parts of their roots each year and in one sense never age. It is impossible to know how old they really are because, unlike trees, they do not produce annual growth rings.

CULTIVATION

Little is known about cultivation of the funnel lily, but it can probably be propagated by planting seed that has been stratified or by transplanting the bulbs. Funnel lilies may need special soil requirements for their growth, because they are often found in rocky areas. However, wild stands of this rare plant should not be exploited.

Apios americana
Groundnut

C. Kuhn
© 86

COMMON NAMES

Groundnut, wild potato, Indian potato, wild sweet potato, American potato bean, wild bean, ground bean, hopniss, Dakota peas, sea vines, pea vines, pomme de terre, and patates en chapelet. (There is much confusion over the common names for groundnuts and for the common peanut, sometimes also called groundnuts; the wild turnip, *Psoralea esculenta* Pursh, also called wild potato and pomme de terre; and the hog peanut, *Amphicarpa bracteata*, also called wild bean, pomme de terre, and pea vines.)

INDIAN NAMES

The Dakota (Sioux) name for the groundnut is "mdo," in the Teton (Sioux) dialect the name is "blo," the Omaha and Ponca name is "nu," the Winnebago name is "tdo," and the Pawnee name is "its" (no translations given) (Gilmore, 1977, p. 42). Although the Cheyenne lived west of the normal range of the groundnut, their name for it was "ai'-is-tom-i-misis'-tuk" (tasteless eating). This same name was given to *Polygonum bistortoides*, found in the Bighorn Mountains of Wyoming (Grinnell, 1962, pp. 173, 179) which may have less flavor than groundnuts.

SCIENTIFIC NAME

A'pios american'a Medic. is a member of the Fabaceae (Bean

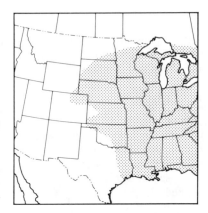

Family). *Apios* is Greek for "pear," referring to the underground tubers of the groundnut, which are sometimes pear-shaped. The species name, *americana*, means "of America."

DESCRIPTION

Perennial herbs from slender rhizomes with tuberous thickenings 1.3–4 cm (½–1⅝ in) thick, stems twining or climbing over other plants. Leaves alternate, divided into 5 to 7 leaflets, egg-shaped, 2–10 cm (¾–4 in) long, sometimes hairy. Flowers in rounded clusters among leaves, from Jul to Sep; petals 5, upper 1 round, white and reddish brown, 2 side wings curved down, brown-purple, lower 2 sickle-shaped, brownish red. Fruits dry, straight or slightly curved, narrow, 5–10 mm (³⁄₁₆–⅜ in) long, two parts coiling after opening; seeds oblong or square, dark brown, surfaces wrinkled.

HABITAT

Wet meadows, low thickets, banks of streams, ponds, sloughs, and moist soil in woodlands.

PARTS USED

Tubers (all year but best in late fall through early spring)—eaten raw or cooked (which is preferable), or dried and ground for flour; seeds (summer)—eaten cooked like peas.

FOOD USE

The groundnut is a common native food plant of temperate, eastern North America. Its distribution reaches west to the wet margins of prairies, where it was once used extensively by the native people. The Reverend J. Owen Dorsey reported in 1881 (p. 308) that Omaha women would dig the "root" in the winter. Also that "there are different kinds of this root, some of which have good skins. Several grow on a common root. These potatos are boiled; then the skins are pulled off, and they are dried." Groundnuts were abundant along the Loup River (in present-day Nebraska), and the Omaha called that river "Nu-tan-ke" "river where 'nu' (groundnut) abounds" (Gilmore, 1913c, p. 323).

Groundnuts were reported to be a source of food among the Dakota and Santee Sioux. In the *Patent Office Report on Agriculture* (1849, pp. 451–452), Philander

Prescott stated that the Dakota, in the "newly sprung up Territory of Minnesota," harvested the roots of "Mendo," or "wild sweet potato." They were found

throughout the valleys of the Mississippi and St. Peters, about the bases of bluffs, in rather moist but soft rich ground. The plant resembles the sweet potato, and the root is similar in taste and growth. It does not grow so large or long as the cultivated sweet potato, but I should have thought it the same, were it not that the wild potato is not affected by the frost. A woman will dig from a peck to a half bushel a day.

The Indians eat them, simply boiled in water, but prefer them cooked with fat meat.

The women of the Prairie Island Santee (Sioux) were reported in 1935 to still be harvesting groundnuts in Minnesota (Landes, 1968, p. 203).

George Bird Grinnell spent many years among the Cheyenne collecting and recording information about their lives (1962, p. 173). He reported that the groundnut was one of their useful plants, but he was unable to obtain a specimen to verify his identification:

From the Indians' description, [I] conjecture that it is this species. The older people speak of red-skinned tubers on the root of a climbing vine, which taste and look like a potato. The rounded leaf is shaped like a teaspoon and somewhat cupped. The plant grows on the North Platte, Missouri, and Laramie rivers. The largest tubers may be the size of a hen's egg. On the vine there may be half a dozen tubers on a single root.

These locations given for the groundnut are further west than any distributions that have been reported in the *Atlas of the Flora of the Great Plains* (Great Plains Flora Association, 1977). It is possible that this plant was propagated by the Cheyenne and other tribes and its range extended westward.

Groundnuts were also important food sources of the Osage and the Pawnee. The Osage gathered them in late summer and fall and stored them in caches for winter use (Matthews, 1961, pp. 443, 478). The Pawnee, with their villages along the Loup and Republican rivers, used groundnuts as a food (Gilmore, 1977, p. 42). Groundnuts were found during the excavation of the Hill site located near what is now Guide Rock in south-central Nebraska (Wedel, 1936, pp. 59, 121). This was a village of the Republican band of the Pawnee during historical times and was probably the one that Zebulon Pike visited in 1806.

The groundnuts that the excavation uncovered were identified by Melvin Gilmore as well as those from four Ozark Bluff-dweller

sites located in Arkansas, which may be the oldest groundnut remains to be found. It was not possible to date these remains, but the Ozark specimens are regarded as pre-Columbian (Beardsley, 1939, p. 512). Gilmore established the Ethnobotanical Laboratory at the University of Michigan, where plant materials obtained from various archeological sites are identified.

Groundnuts were also an important food of New England colonists. Alexander Young, in his *Chronicles of The Pilgrim Fathers* (1841, in Hedrick, 1919, p. 55) reported that during their first winter, the Pilgrims "were enforced to live on ground nuts." The great value of groundnuts to the colonists as a food is indicated by the fact that there reputedly was a town law in 1654 stating that if an Indian dug groundnuts on English land, he was to be set in stocks; for a second offense he was to be whipped (Fernald et al., 1958, p. 254).

Peter Kalm, an economics professor and botanist from Sweden, observed the hopniss (groundnut) when in the northern colonies. He reported on March 17, 1749:

Hopniss or hapniss was the Indian name of a wild plant which they ate . . . and it grows in the meadows in a good soil. The roots resemble potatoes, and were boiled by the Indians, who eat them instead of bread. Some of the Swedes at that time likewise ate this root for want of bread. Some of the English still eat them instead of potatoes. Mr. Bartram told me that the Indians who live farther in the country do not only eat these roots, which are equal in goodness to potatoes, but likewise take the pease which lie in the pods of this plant, and prepare them like common pease (Pinkerton, 1812, p. 533).

Groundnuts that I have grown in my garden in northeast Kansas and those that I have found growing wild in our region seldom set seeds.

There has been considerable confusion over the identification of groundnuts by explorers of the Prairie Bioregion. Edward Palmer in *Report of the U.S. Commission of Agriculture* (1871, p. 405) stated that "the tuber of this common plant, which grows on the banks of streams and in alluvial bottoms, is the true 'pomme de terre' of the French." This name means "fruit of the earth" and most commonly refers to the prairie turnip, *Psoralea esculenta* Pursh, but was used as a general term for any edible food from the earth.

Edwin James, botanist for the Long Expedition, is responsible in part for some of the confusion surrounding the identification and names of the groundnut. In 1819, while at the Omaha village he reported that: "The squaws, . . . are often necessitated to dig the pomme de terre, 'noo'; and to scratch the groundpea, 'himbaringa,' (the same word is also applied to the bean) from beneath the soil. This vegetable is produced on the roots of the apios tuberosa, they also frequently find it hoarded up in the quantity of a peck or more in the brumal retreats of the field mouse for its winter store" (Thwaites, 1905, 14: 309). It appears that James was confusing the groundnut (*Apios tuberosa*), which the Omaha called "noo" or "nu," with the hog peanut (*Amphicarpa bracteata*), which the Omaha called "hinbthihi" or "himbaringa." It is most likely that the seeds gathered by field mice were those of the hog peanut.

While heading up the Missouri River on July 10, 1804, the Lewis and Clark expedition stopped to "dine" on Solomon's Island (near what is now the intersection of the Nebraska, Missouri, and Kansas borders). Just to the west of the river was "a butifull bottom Plain of about 2000 acres covered with wild rye & Potatoes, intermix't with the grass" (Thwaites, 1904, 1: 73). Thwaites, the editor of Lewis and Clark's Journal, calls these potatoes "ground apple pomme de terre" and in a footnote identifies them as the prairie turnip, *Psoralea esculenta*. I believe that they were groundnuts because of their location (prairie turnips are not usually found in bottomland with tall grasses) and because of references that follow to similar habitat locations for groundnuts.

Captain L. C. Easton traveled from Fort Laramie (along the Platte River in Wyoming) to Fort Leavenworth in 1849 on an assignment to find a more direct supply route to the Oregon Trail. On September 10, he made reference to finding an abundance of "Sea Vines" (probably groundnuts) in present-day northeast Kansas near Manhattan: "We found a large quantity of Sea Vines on the River at our present encampment, and our Animals appeared to enjoy them exceedingly. This Vine is plentiful on all the Creeks, from this point to Fort Leavenworth—It is a fine food for Horses and Mules" (Mattes, 1952, p. 410). The groundnut, being a member of the legume family, would be expected to produce a high-protein feed from its leaves and stems. Apparently, this plant is enjoyed by livestock, because it has been grazed out of most of its former locations in pastures.

Groundnut tubers are a good source of carbohydrates and contain between 13 and 17 percent protein by dry weight, or about three times more than potatoes or any other widely used vegetable root (Yanovsky and Kingsbury, 1938, p. 657; Walter et al., 1986, p. 40; Watt and Merrill, 1963, p. 106). The tubers are edible raw, but have a milky juice and are slightly rubbery tasting. They are best when boiled for a few min-

utes. When prepared in this manner, they have a more earthy, rich taste than potatoes, and even children to whom I have served them have described them as "delicious."

CULTIVATION

The botanist Constantine Rafinesque (1783–1840) reported that groundnuts were cultivated by the Creek Indians, not only for their tubers, but also for their seeds, which he said were as good as peas (Havard, 1895, p. 101). It is quite likely that groundnuts were one of the plants that was semicultivated by the Indians of the prairies. Tubers could be easily propagated by carrying them to another place and planting them in a suitable habitat. This sort of semicultivation may have been practiced by the Cheyenne, who reported that groundnuts grew along the North Platte and Laramie Rivers, several hundred miles west of their native range (Grinnell, 1962, p. 169). Also, I have observed groundnuts growing in a seepy, marshy area directly below a petroglyph (Indian rock carving) site in central Kansas (Ellsworth County). It is possible that they were planted there by Indians.

Asa Gray, an eminent American botanist, stated in 1874 that "our Ground-Nut would have been the first developed esculent tuber and would probably have held its place in the first rank along with potatoes and sweet potatoes of later acquisition" had our civilization started in America instead of Asia and Europe (Havard, 1895, p. 102). Groundnuts were cultivated in France as early as 1635, but were soon forgotten. They were reintroduced into Europe in 1845, this time as a substitute for potatoes, which had become subject to disease. However, groundnuts never proved to be a practicable crop (Fernald et al., 1958, p. 254).

Dr. V. Havard of the U.S. Army reported (1895, p. 102) on some experiments and observations relating to groundnut cultivation: "Native cultivation does not appear to have had any effect upon the size and quality of this tuber, and that experiments by Vilmorin and others with a view to its improvement have not been successful, although hardly continued long enough to be conclusive. The tuber is of slow growth, requiring two or three years before reaching sufficient size to be useful, and its creeping, scattering habit renders the harvest laborious."

I have successfully grown groundnuts in river-bottom ground in Lawrence, Kansas. Tubers can be planted two to three inches deep in the early spring. After shoots come up, they should be mulched to stop competition from weeds and grass, and the vines should be provided with something upon which to climb. After one year of growth, several one inch-thick tubers can be harvested from each plant. Because of their vining nature, groundnuts would be hard to grow on a field scale, and their annual yields appear to be quite low in comparison to other crops. Also, they would be difficult to cultivate mechanically, because each tuber can sprout and grow in the spring, filling in spaces between rows.

There is a second species native to eastern North America, *A. priceana* Robbins, which is found in distinct habitats in Alabama, Mississippi, Tennessee, Kentucky, and Illinois (Walter et al., 1986, p. 39). It has a single, irregular, turnip-shaped root six inches or so in diameter (Fernald et al., 1958, p. 255). Selective breeding of these two species could lead to the development of a new agricultural crop. Research at Louisiana State University is offering new hope for the groundnut as a future food crop. By using mass selection and genetic manipulations, yields of over eight pounds of tubers per plant have been obtained (Bill Blackmon, 1986, personal communication). Blackmon also has started an informative newsletter called "The Apios Tribune" as a forum for the renewed interest in groundnuts.

The groundnut is also an attractive ornamental. It is described in the 1882 *Canadian Horticulturist*:

This is a little gem; in July or August it is one mass of chocolate colored, pea shaped flowers, which is a very unusual color in flowers. Its leaflets are very pretty also. It grows upon the low bushes of the Northern woods and often lends beauty to a hazel bush, which is rarely very fine itself. . . . [It] can be grown in a window, and would be a fine ornament if the tubers were started late in summer so as to throw its flowering season late in the fall, but as a garden climber it would be fine planted amongst tall-growing summer roses, as it would do them no harm, but lend a beauty to them after they had done blooming (Erichsen-Brown, 1979, p. 371).

Asclepias syriaca
Common Milkweed

C. Kuhn
© 86

Common milkweed, wild asparagus, milk plant, silkweed, virginia
silk, and wild cotton.

INDIAN NAMES

The following names of common
milkweed do not have translations: Pawnee, "kari'piku";
Omaha and Ponca, "wahtha"; and
Winnebago, "mahin'tsh" (Gilmore, 1977, p. 57). The Dakota
name for the showy milkweed,
Asclepias speciosa, is "pan-
nun'pala" (two little workbags of
women), perhaps referring to the
shape of the pod (Rogers, 1980a, p.
47). The Dakota and Lakota also
call it "waxca-xca" (flower blossom) because they used the flower
blossoms for food (Munson, 1981,
p. 232). The Cheyenne called this
plant "ma-tan-ai'-mahdst" (milk,
pieces of wood) (Grinnell, 1962, p.
184).

SCIENTIFIC NAME

Asclep'ias syria'ca L. is a member
of the Asclepiadaceae (Milkweed
Family). *Asclepias* comes from
the name of the Greek god of
medicine, Asklepios. The species
name *syriaca* means Syrian, an
odd misnomer for this native
North American species.

DESCRIPTION

Perennial herbs with erect stems
6–20 dm (24–80 in) tall, hairy,
containing milky white juice.
Leaves opposite, oval to elliptic or

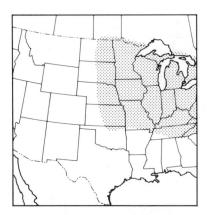

oblong, 10–20 cm (4–8 in) long,
hairy. Flowers small, in rounded
clusters at ends of stalks among
upper leaves, from May to Aug;
petals 5, bent downward, pinkish
to purple, topped by a crown of 5
erect lobes, pinkish to white.
Fruits dry, spindle-shaped, 8–10
cm (3–4 in) long, surfaces covered
with small, knoblike projections,
brown, opening to release seeds
bearing tufts of white, silky hairs.

A. speciosa Torr. is very similar,
with only minor differences in
structure of the flowers, but grows
mainly in the western Great
Plains, whereas *A. syriaca* is more
common in the eastern part.

HABITAT

Prairies, pastures, and old fields.

PARTS USED

Young shoots (spring)—cooked as
an asparagus substitute; flowers,
buds, and immature fruits (late
spring to summer)—cooked; dew-

covered flowers reported to yield sugar. Note that all parts of this plant have some poisonous properties and should be cooked for four minutes, with at least one change of water, to remove the toxins before eating.

The Omaha ate the common milkweed during three stages of its growth—"the young sprouts in early spring, like asparagus sprouts; the clusters of floral buds; and the young fruits while firm and green. It is prepared by boiling" (Gilmore, 1977, p. 57). When the Omaha and Pawnee first saw cabbage and noted that it was boiled as they boiled milkweed, they called cabbage "White man's milkweed" (ibid.). The Osage also compared cabbage to milkweed and harvested the young milkweed sprouts and floral buds. Sometimes they stayed on the prairies during their western hunting and gathering trips long enough to eat the delicious young fruit pods (Matthews, 1961, p. 457).

The Dakota and Lakota are closely related and speak similar Siouan languages. Both of these tribes used showy milkweed as food (Munson, 1981, p. 232; Rogers, 1980a, p. 47), and the Dakota were also reported to use the common milkweed (Gilmore, 1913b, p. 363). John Frémont reported on June 22, 1842, that he found the "milk plant" (*A. syriaca*) in con-

siderable quantities along the Blue River in southeast Nebraska, and stated: "The Sioux Indians of the Upper Platte eat the young pods of this plant, boiling them with the meat of buffalo" (Jackson and Spence, 1970, p. 180).

The Crow Indians boiled the flowers of showy milkweed for food and ate the seeds raw, when they were young and tender in the immature fruits (Blankenship, 1905, p. 7). The Cheyenne also ate the flower buds of showy milkweed, which they boiled in water, often with meat, grease, gravy, or soup. The liquid resulting from boiling the buds was flavored with the "layer underneath the hair of buffalo or deer hides" (Hart, 1981, p. 14). The young fruits of milkweed (species not identified) were cooked and eaten by the Kiowa after the hairy outer surface had been removed. Also, the dried pods served as spoons (Vestal and Schultes, 1939, p. 48).

Edward Palmer (1871, p. 405) reported on the use of the butterfly milkweed, *A. tuberosa* L.: "The stem of this plant expands under ground into a tuber of considerable size, which is boiled for food. The flowers are odoriferous, and the Sioux of the Upper Platte prepare from them a crude sugar by gathering them in the morning before the dew is evaporated. They also eat the young seed-pods of the plant, after boiling them with buffalo meat. Some of the Indians of Canada use the tender shoots of this species as we use asparagus."

One of the problems of older accounts is that plant species often are not clearly identified. In this case, it appears that Palmer may have confused butterfly milkweed with common milkweed, because butterfly milkweed only occurs in the easternmost portion of the Sioux territory. Also, part of the information that he gives apparently comes from John Frémont (quoted above) and part comes from Peter Kalm (which follows) and both of them identify the plant as common milkweed.

Peter Kalm, a Swedish professor of economics and a botanist, traveled widely in the American colonies to study the native vegetation and find plants of economic value that might be grown in Sweden. On July 12, 1749, near Fort St. Frederic, along the eastern shore of Lake Champlain (in Vermont), he wrote about the common milkweed:

When the stalk is cut or broken it emits a lactescent juice, and for this reason the plant is reckoned in some degree poisonous. The French in Canada nevertheless use its tender shoots in spring, preparing them like asparagus, and the use of them is not attended with any bad consequences, as the slender shoots have not yet had time to suck up anything poisonous. Its flowers are very fragrant, and when in season, they fill the woods with their sweet exhalations and make it agreeable to travel in them, especially in the evenings. The French in Canada make a sugar of the flowers, which for that purpose are gathered in the morning, when they are covered with dew. This dew is pressed out, and by boiling yields a very good brown, palatable sugar (Pinkerton, 1812, p. 612).

The common and showy milkweeds are my favorite wild foods from the prairie. Milkweed was the wild food I harvested most frequently on my walk from Kansas City to the Rocky Mountains during the summer of 1983. I prepared seven meals with common milkweed and one with showy milkweed. I observed that common milkweed is a "weedier" species and much easier to find on roadsides and field margins than is showy milkweed, which seems to be more abundant in prairie habitats.

I also sampled cooked butterfly weed in Ellsworth County in central Kansas. It is not recommended as an edible prairie plant; it is quite hairy and is a fairly uncommon and attractive wildflower. Three weeks later, in Wallace County in western Kansas, I sampled cooked broadleafed milkweed, *A. latifolia* (Torr.) Raf., which is known to be poisonous to livestock. It tasted good, and I felt fine after eating it, but because of possible problems with toxins, I cannot recommend it either.

Milkweed does contain bitter-tasting toxins, which can be removed by boiling for four minutes and then changing the water (Gaertner, 1979, p. 122). There have been reports of pain and discomfort when this procedure has not been followed (ibid.). Thomas Elias and Peter Dykeman, in *Field Guide to North American Wild Edible Plants* (1982, p. 106), suggest changing the water two or three times. Some species of milkweed (such as broadleafed milkweed mentioned above) are known to have greater quantities of the toxins, which include cardiac glycosides, resinoids, and a few alkaloids (Stephens, 1980, pp. 85, 86).

The common and showy milkweed also can be confused with dogbane, *Apocynum cannabinum* L., which is somewhat poisonous and more toxic than milkweed. Milkweed shoots, like asparagus, pop out of the ground quickly in the spring and are hard to find unless the location of a favorite milkweed patch is known. It is possible to confuse the young shoots of dogbane with those of common and showy milkweed because they also have milky sap and opposite leaves and come up at the same time. However, the young stems of dogbane are smooth, tough, sometimes reddish, and not usually as fat or succulent as those of milkweed and they quickly fork. I have tasted young dogbane shoots cooked and found them to be quite bitter—and as a rule food that does not

taste good probably is not good for you.

Milkweed has many other potential uses. The dried milky sap has been used as a chewing gum (Fernald et al., 1958). The plant was experimented with as a substitute for latex (Gaertner, 1979), as fuel plant for a "petrochemical plantation" (Nielson, 1977), and as a fiber plant (using the parachuting seed-hairs that many children play with in the fall) (Gaertner, 1979). Renewed interest in these uses for milkweed has prompted Kansas State University to begin developing a hybrid that could be grown as an alternative crop in the Midwest.

CULTIVATION

The common milkweed may have been cultivated, or at least encouraged by some Indian tribes. Huron Smith in *Ethnobotany of the Potawatomi* (1933, p. 42) reported: "One always finds a riot of milkweed close to the wigwam or house of the Indian, suggesting that they have been cultivated." Of course, this could be attributed to the milkweeds invading a disturbed habitat or children playing with the seeds. One of the first reports of milkweed being cultivated as a garden crop appeared, along with an unusual use for the silky seed hairs, in the 1633 Gerarde-Johnson Herbal, published in London: "This plant . . . . is kept in some gardens by the name of Virginia Silke Grasse. . . . The silke is used of the people of Pomeioc and other of the Provinces adjoyning, being parts of Virginia, to cover the secret parts of maidens that never tasted man" (Erichsen-Brown, 1979, p. 432).

Common and showy milkweeds can easily be established in a wildflower garden or as part of a semiwild patch of prairie plants. They are easily grown from spring cuttings or root division in the fall or early spring. Seeds can be planted in fall or stratified and planted in the spring.

Astragalus crassicarpus
Groundplum Milkvetch

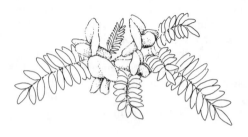

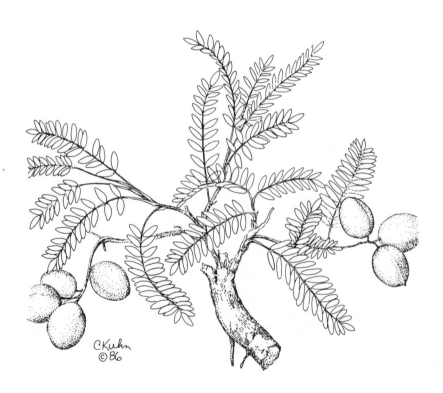

CKuhn
© 86

COMMON NAMES

Groundplum milkvetch, ground plum, buffalo bean, buffalo pea, and Indian pea.

INDIAN NAMES

The Dakota name is "pte ta wote" (buffalo food) (Gilmore, 1977, p. 40). The Omaha and Ponca had two names for this plant—"tdika shande" and "wamide wenigthe" (something to go with seed) because they used it for preparing corn seed for planting (ibid.).

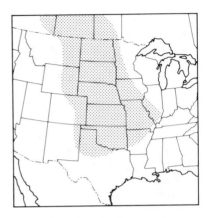

SCIENTIFIC NAME

Astrag'alus crassicar'pus Nutt. is a member of the Fabaceae (Bean Family). *Astragalus* comes from the ancient Greek name for a plant in the pea family. The species name *crassicarpus* means "thick fruit," referring to the thickened plumlike pods.

DESCRIPTION

Perennial herbs from thick, woody taproots; several stems usually lying on the ground, 1–6 dm (4–24 in) long, hairy. Leaves alternate, pinnately compound, 4–13 cm (1⅝–5 in) long, with 15–27 leaflets, elliptic to oblong, lower surfaces hairy. Flowers in elongated groups among leaves, from Mar to Jun; petals 5, upper 1 large and erect, lower 2 boat-shaped, 2 wings at sides, purple, light blue, or rarely whitish. Fruits fleshy, becoming dry, round or oblong, 1.5–2.5 cm (⅝–1 in) long, smooth,

greenish turning red or purple where exposed to sun; seeds black.

HABITAT

Prairies, pastures, limestone outcroppings, and rocky open woods.

PARTS USED

Immature pods (spring)—raw, cooked, or pickled.

FOOD USE

In the early spring, the purple-pink flowers of the groundplum milkvetch are a colorful indicator that the prairies are beginning their yearly renewal of growth. These flowers are soon followed by succulent, edible, plumlike fruits that sprawl on the ground. They taste like a sweet, watery, green pea and are a good snack food during wildflower walks across a prairie. These fruits were a minor food source of the original prairie inhabitants.

The Dakota sometimes ate the fruits of the groundplum milkvetch fresh off the plant (Gilmore, 1913b, p. 365). The Pawnee, when traveling in the early summer, often picked these fruits and ate them to prevent thirst (Dunbar, 1880, p. 324). J. W. Blankenship, botanist for the Montana Agricultural Experiment Station, reported in "Native Economic Plants of Montana" (1905, p. 7) that "the fleshy, plum-like pods have a sweetish insipid taste and were eaten by the various tribes of Indians in July both raw and boiled. Now they are occasionally used for pickles."

Both the Omaha and Ponca previously gathered the fruits of this plant, which are formed at corn-planting time, put them with their seed corn, and soaked them both in water. When the corn seed had sufficiently soaked to start germinating, the groundplum milkvetch fruits were thrown away and the corn seed was planted (Gilmore, 1977, p. 40). This was an old custom, the origin and meaning of which have been forgotten. One apparent benefit of this practice is that by using the groundplum milkvetch as a seasonal indicator, the corn planting schedule was attuned to the yearly climate pattern. If it were a cold, wet spring, the groundplum milkvetch would bloom and set fruits later than normal, thus appropriately delaying the time to soak and plant corn seed.

The groundplum milkvetch is a characteristic plant of well-drained upland prairies, where sometimes it is the most important nonwoody plant other than grass. This low-lying plant grows rapidly from its crown and coarse taproot in the spring, so that it can bloom and set seeds before it is shaded by the taller grasses. The size of the fruit is a good indicator of the availability of soil moisture in a specific location. In areas of lighter rainfall, where there is over-grazing, or during times of drought, the fruits (and the entire plants) may attain only two-thirds or even one-half the size of those in prairies with a better water supply (Weaver and Fitzpatrick, 1934, p. 210).

Positive identification of groundplum milkvetch is necessary if it is to be used as a food because there are several closely related species called poison milkvetch or locoweed—*Astragalus racemosus, A. bisulcatus, A. pectinatus,* and *A. mollissimus; Oxytropis lambertii* and *O. sericea.* These locoweeds produce a toxic alkaloid substance, accumulate selenium when growing on selenium-bearing soils, or both. Poisoning by these plants has been known to make cattle and horses "loco" (Stephens, 1980, pp. 46, 47; Bare, 1979, p. 165). To be safe, any plant that resembles groundplum milkvetch and is growing on alkali flats or other areas that may have selenium-bearing soils should not be harvested. Such areas are often devoid of other vegetation. Also, groundplum milkvetch can be distinguished from its poisonous relatives by its uniquely rounded, large, fleshy fruits.

CULTIVATION

Groundplum milkvetch can be grown in a wildflower garden, most easily from stratified seed, or by root division from mature plants in the spring.

Atriplex subspicata
Saltbush

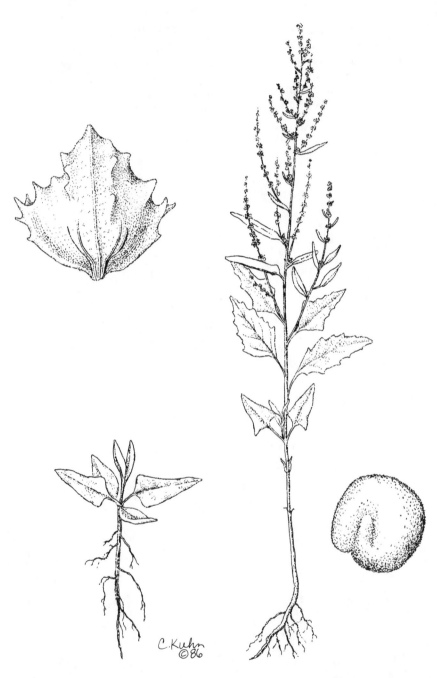

C. Kuhn
©86

COMMON NAMES

Saltbush, spearscale, and orache.

INDIAN NAMES

None were found in the sources consulted.

SCIENTIFIC NAME

A'triplex subspica'ta (Nutt.) Rydb. is a member of the Chenopodiaceae (Goosefoot Family). *Atriplex* probably means "black or dark network," referring to the conspicuous veination pattern on the small leaflike appendages beneath the flowers. The species name *subspicata* means nearly spikelike, referring to the arrangement of flowers.

DESCRIPTION

Annual herbs, 3–10 dm (12–40 in) tall, often much branched, stems angular, with green to reddish stripes. Leaves opposite near base, alternate above, lance-shaped or oblong, 3–12 cm (1¼–4¾ in) long, fleshy, often with a pair of blunt lobes near bases, margins sometimes irregularly toothed. Flowers in compact clusters among leaves on upper branches, from Jun to Aug; male and female separate, greenish with 2–5 segments. Fruits dry, enclosed by leafy structures, each containing 1 seed, brown and dull, 1.5–3 mm (¹⁄₁₆–⅛ in) wide or black and shiny, 1–2 mm (±¹⁄₁₆ in) wide.

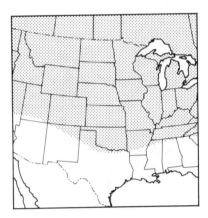

HABITAT

Alkali salt flats; cultivated and waste ground.

PARTS USED

Greens and shoots (spring to fall)—cooked; seeds (fall)—raw, cooked, or for tea; ashes of the plant for leavening.

FOOD USE

Saltbush has a salty taste and grows on saline soils. While there is no record of the use of saltbush by Indians of the Prairie Bioregion, it was almost certainly a source of food. This supposition is based on the fact that at least 13 species of *Atriplex* were used as food and flavoring by Indians of the Southwest and Great Basin (Yanovsky, 1936, p. 21). Some of those species (mentioned below) extend into the prairies, especially the dry and alkali areas of the shortgrass prairie. They can be divided into two groups—perennial woody shrubs and annual non-woody plants.

Frank Russell reported in 1908 that the Pima (of southern Arizona) used the shrubby, perennial saltbush, *A. nuttallii* S. Wats.: "The young stems and flower heads are boiled with wheat for the purpose of flavoring, the stems being cut in short lengths and sometimes used as a stuffing in cooking rabbits" (Castetter, 1935, p. 18).

Other species of saltbush, especially *A. argentea* Nutt. and *A. cornuta* M. E. Jones, were eaten by the Pueblo Indians of the Rio Grande Valley. They were boiled and eaten alone or after boiling were included with various plants and meats for flavoring. The Isleta also ate *A. argentea* as cooked greens and the people at the Acoma and Laguna pueblos used it for food and as a salty flavoring (Castetter, 1935, p. 18). The Hopi use the ashes of fourwing salt-bush, *A. canescens* (Pursh) Nutt., as a substitute for baking powder (no description of their method was given) (Krochmal et al., 1954, p. 12).

The Hopi, as well as the Cochiti, Acoma, and Laguna Pueblo Indians ate the salty leaves of *A. powellii* S. Wats. as cooked greens. This plant was the earliest one harvested of all the Hopi spring food plants (Castetter, 1935, p. 18). The Zuni said that before their people had corn, the seeds of this saltbush and other species were eaten raw. After their people had corn, the seeds were ground with corn meal and made into mush, and they said: "When we depended entirely on the small seeds of plants for our food, our flesh was not firm and good as it is now" (Stevenson, 1915, p. 66).

Saltbush has the ability to tolerate the severe environmental conditions found in saline soils and alkali flats. As it grows, it apparently assimilates some of the salts. Saltbush species also accumulate selenium, a beneficial mineral for humans when consumed in small amounts, but poisonous in large doses. In the areas where selenium is known to occur in the soil, saltbush should be eaten in moderation. When eaten as young greens, saltbush tastes like a presalted lamb's quarters, to which it is related. Older plants may be tough and have an unappetizing taste.

This same species is also found growing along the Atlantic seashore. In *Edible Wild Plants of Eastern North America*, Fernald et al. (1958, p. 182) reported that the succulent leaves and young tips of plants, found along the seashore and used as a pot herb, taste "superior to Lamb's-Quarter or Pigweed."

CULTIVATION

Saltbush was encouraged by Indians in the Great Basin, who burned the native vegetation at the appropriate time of the year to promote its growth (Winter, 1974, p. 121). One species of saltbush, *A. hortensis* L., also called orach or garden orach, is an annual garden crop. This native of Asia is grown for its edible leaves and ornamental foliage, which varies from yellow to green to red. Our native annual saltbush, *A. subspicata*, can be grown from seed. However, it and other native species of *Atriplex* have pollen that is known to cause hay fever (Bare, 1979, p. 87).

Callirhoe involucrata
Purple Poppy Mallow

C. Kuhn
©86

Purple poppy mallow, poppy mallow, low poppymallow, and purple mallow.

INDIAN NAME

The Lakota name for this plant is "ezhuta nantiazilia" (medicine or smoke treatment medicine), referring to its use for producing a medicinal smoke (Gilmore, 1977, p. 51).

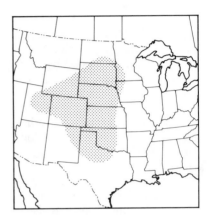

SCIENTIFIC NAME

Callirho'e involucra'ta (T. & G.) Gray, is a member of the Malvaceae (Mallow Family). The genus *Callirhoe* refers to an ocean nymph in Greek mythology and comes from the Greek "kallirrhoos," meaning "beautiful-flowering." The species name *involucrata* refers to the involucre, a whorl of leaflike structures just below the flower.

DESCRIPTION

Perennial herbs from elongated roots, stems mostly trailing, to 7 dm (28 in) long, hairy. Leaves alternate, rounded, 2.5–7 cm (1–2¾ in) long, divided into 5–7 parts, each one toothed, lobed, or deeply cut. Flowers solitary among upper leaves, on stalks longer than leaves, from Feb to Aug; petals 5, separate, 1.5–3 cm (⅝–1¼ in) long, rose to purple, prominent columns of yellow stamens in centers. Fruits dry, round and flattened, 3–5 mm (⅛–³⁄₁₆ in) tall, splitting into 2 parts.

HABITAT

Prairies, pastures, fields, roadsides, grassy open areas, and disturbed soil.

PARTS USED

Roots (fall or early spring)— cooked; leaves (spring, summer)— to thicken soup.

FOOD USE

This showy perennial plant, with its fragrant and beautiful summer flowers, develops a sweet starchy root that tastes somewhat like a sweet potato. The root is about the size of a carrot and is harvested in late summer or fall. The Osage dug the roots of purple poppy mallow and stored them in caches to eat during the winter (Matthews, 1961, p. 443).

In 1849, Howard Stansbury of

the Corps of Topographical Engineers of the United States Government was traveling on the Oregon Trail through Nebraska on his way to Salt Lake City. On June 24, 56 miles west of Fort Kearney, he described seeing the "purple mallow (the root of which resembles the parsnip, and it is used by the Indians for food)" (McKelvey, 1955, p. 1081).

It is surprising that poppy mallow has seldom been mentioned as an Indian food source. It also has not been positively identified in archaeological remains, but this could be attributed to the fact that fleshy plant materials do not preserve well. Seeds of an unidentified mallow species have been found in northwestern New Mexico in Salmon Ruin, an Anasazi Pueblo site that was occupied from the eleventh to the thirteenth century A.D. However, it is not known for what purpose these seeds were used.

The Long Expedition originally set out in 1819 to explore the Yellowstone River, but because of a huge fiasco involving political corruption linked to the expedition, funds were cut by Congress. Plans were changed and the expedition, with only the bare essentials, set out instead from Council Bluffs (now a city in Iowa across the Missouri River from Omaha) to explore the source of the Platte River. Dr. E. James, the botanist on the expedition, was apparently describing poppy mallow when he reported: "In the fertile grounds,

along the valley of the Loup fork (in present-day Nebraska), we observed several plants which we had not seen . . . one belonging to the family of the *Malvaceae*, with a large tuberous root which is soft and edible, being by no means ungrateful to the taste" (McKelvey, 1955, p. 214).

Lieutenant James William Abert, while on a topographical assignment to examine the route from Fort Leavenworth (in northeast Kansas) to Bent's Fort (in eastern Colorado), also noticed and later reported on some natural history of the region. On July 8, 1846, three days west of Council Grove (in central Kansas) he recorded that "we found some beautiful plants with brilliant scarlet flowers (*Malva pedata*) and roots which are eatable." (McKelvey, 1955, p. 985). This plant was the purple poppy mallow, because the name *M. pedata* is now recognized as a synonym for *Callirhoe involucrata*.

Dr. V. Havard of the U.S. Army reported (1895, p. 111) that the genus *Callirhoe* is the only one in the Mallow Family with fleshy, edible roots and that a species closely related to purple poppy mallow, the fringed poppy mallow, *C. digitata* Nutt., of the southern plains, has a root "in shape and size between a small turnip and a parsnip, said to be even more pleasant tasted than that of *Psoralea* (the prairie turnip) and highly prized by the natives."

The leaves of purple poppy mallow are also edible and have a pleasant taste. Like okra pods and the leaves of some other members of the Mallow Family, they are mucilaginous and are good for thickening soup. This use for mallows seems to be almost universal; a family with whom I gardened in Columbia, Missouri, had brought with them from Korea seeds of an annual mallow, which they grew for its leaves to thicken soup.

CULTIVATION

Purple poppy mallow is easy to grow in a sunny location and, like most prairie plants, is rather drought-tolerant. It can be started either from seeds or from a root cutting. In the fall of 1981, I transplanted a root from my parents' farm in southern Nebraska to my garden in the alluvial soil of North Lawrence, Kansas. During the summer, it grew rapidly and formed a circular mat, about four feet across, that had numerous crimson-red flowers. On a warm day, especially after a rain, the fragrance of the flowers filled the air with a sweet perfume, and this scent in combination with the brilliant color of the flowers, attracted bumblebees. In the fall, I harvested a parsnip-sized root and replanted a smaller piece from the crown. It has continued growing ever since. Purple poppy mallow should be considered as an addition to gardens for its tasty, cooked roots, its soup-thickening leaves, and its attractive flowers.

Camassia scilloides
Wild Hyacinth

C. Kuhn
© 86

Wild hyacinth, Atlantic camassia, camass, eastern camass, wild camas, camass lily, meadow hyacinth, and meadow quill.

INDIAN NAMES

The Comanche name is "siko" (no translation given) (Carlson and Jones, 1939, p. 520).

SCIENTIFIC NAME

Camas'sia scilloi'des (Raf.) Cory., is a member of the Liliaceae (Lily Family). *Camassia* comes from the Indian name "kamas," meaning sweet, in reference to the baked roots from the similar-looking wild camas found in the northwestern United States. The species name *scilloides* means similar to scilla (a bulbous plant of the Mediterranean).

DESCRIPTION

Perennial herbs up to 7 dm (28 in) tall, growing from bulbs. Leaves basal, linear, 4–20 mm (³⁄₁₆–³⁄₄ in) wide, with a central ridge, shorter than flowering stalks. Flowers in elongated groups on erect stalks, from Apr to Jun; each with 6 separate segments, 7–15 mm (¼–⅝ in) long, pale blue, violet, or white. Fruits dry, roundish, opening into 3 sections; seeds black.

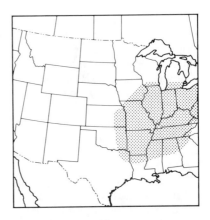

HABITAT

Prairies, meadows (in rich damp soil), rocky (often limestone) glades, and open woods.

PARTS USED

Roots—raw or cooked (late spring, summer, before the tops disappear).

FOOD USE

Wild hyacinth has an edible bulb, but because it closely resembles the poisonous death camas, *Zygadenus nuttallii* Gray, its use as an edible plant is not recommended unless one is absolutely certain of its identification. Wild hyacinth is closely related to wild camas, *C. quamash* (Pursh) Greene, which was the most widely used food root of the Indians of intermountain and coastal regions of the Northwest United States and British Columbia.

During the summer of 1933, some older, non-English speaking

members of the Comanche tribe were interviewed on their reservation in Oklahoma concerning their use of plants. They reported that wild hyacinth was a food source and that they ate the roots raw (Carlson and Jones, 1939, p. 520). Other references to wild hyacinth use cannot be found. Fernald, Kinsey, and Rollins state in *Edible Wild Plants of Eastern North America* (1958, p. 133) that "it is possible that the more eastern species, *Camassia scilloides* is edible, but we have found no evidence of its being eaten."

I have dug the roots of wild hyacinth in June along the woody border of a northeast Kansas prairie remnant. When eaten raw, they tasted starchy, crunchy, and slightly sweet. When boiled for 20 to 30 minutes or baked for 45 minutes at 350°F, the roots are pleasant tasting, but gummy textured (Peterson, 1978, p. 136).

Although it takes a lot of time and effort, the Indian pit method of preparation produces bulbs that are very dark and sweet. This method, commonly used for wild camas, involves the following steps. Dig a big pit (2 feet by 3 feet and about 2 feet deep). Line the pit with stones and build an intense fire inside the pit to heat the stones and adjacent earth. After several hours, clean out the burning wood and ashes, line the rocks with green grass or leafy twigs, and put in the bulbs to be roasted. Cover with more grass and pour in a pint of water to produce steam.

Top the pit with two inches of soil and build a fire on top. To obtain the sweetest mass of baked wild camas bulbs, the Nez Perce, Flathead, and Blackfoot would maintain the fire on top of the pit for 12 to 70 hours (Elias and Dykeman, 1982, pp. 14, 65; Hart, 1976, p. 17).

Pit baking was used because the long period of heat exposure apparently converted a greater percentage of the starchy and nondigestible inulin (the major form of carbohydrate in *Camassia*) to sugar. One reason that the wild hyacinth may not have been a more popular food source for Indian tribes was that they were well aware of the fact that inulin-containing foods produced flatulence.

Another factor that could have limited the use of wild hyacinth bulbs was the possibility of their confusion with the poisonous death camas. William Chase Stevens in *Kansas Wild Flowers* (1961, p. 37) reported that "all parts of death camas are poisonous to man ... in early spring, when the basal leaves are up, there is so strong a resemblance between them (death camas) and the leaves of the wild hyacinth and wild onion that even the Indians when digging bulbs of the 2 latter for food sometimes included by mistake bulbs of the death camas and were seriously poisoned by them."

Death camas and wild hyacinth usually can be distinguished from each other by their flowers: wild hyacinth has blue flowers, whereas death camas flowers are white, green, or bronze (however, white-flowered wild hyacinths are occasionally found). Moreover, when cross sections of the leaves are cut, the midrib in the wild hyacinth is seen to be hollow, whereas that of death camas is solid.

I was curious about the similarity in taste between death camas and wild hyacinth, and although I would discourage anyone from doing the same, I tasted the root of death camas. As might be expected, it was extremely bitter and tasted poisonous. This has led me to believe that confusion between the two plants, and subsequent poisoning, would probably occur only if a few death camas roots were inadvertently added to a strong-tasting stew or baked with a large number of wild hyacinth roots; or if people who mistakenly believed that all wild foods are beneficial were trying to prove their ability to "live off the wild" and ate death camas roots, even if they did taste bad.

CULTIVATION

Wild hyacinth—a beautiful light blue wildflower in the Lily Family—can be cultivated as an attractive ornamental and as a semi-wild food. It is not generally common across its range but sometimes it can be found in large colonies. To ensure the continued health of wild populations, wild hyacinth should be cultivated, rather than harvested in the wild. Camassias are hardy and do well in loamy soils and can be propagated by seeds, although they may not sprout the first year. For ornamental purposes, it is recommended that the bulbs be planted 3 to 4 inches apart in early autumn (Bailey, 1976, p. 208).

Julian Steyermark (1981, p. 465) reported a form of wild hyacinth, C. scilloides f. Petersenii, that is found only in two counties of eastern Missouri growing on limestone slopes. It has roundish bulbs two inches across, which are twice as large as normal. Perhaps with some plant breeding, an edible cultivar of wild hyacinth could be developed.

Ceanothus americanus
New Jersey Tea

C. Kuhn
©86

New Jersey tea, redroot, ceanothus, inland ceanothus, Indian
tea, soapbloom, wild snowball,
wild lilac, mountain sweet, buck-
brush, and snowbrush.

INDIAN NAMES

The Omaha and Ponca name is
"tabe-hi'" (no translation given)
(Gilmore, 1977, p. 50). The Lakota
name for the small redroot, *Ceanothus herbaceus* Raf. var. *pubescens* (T. & G.) Shinners, is "upan'
tawo'te" (female elk food) (Rogers,
1980a, p. 87).

SCIENTIFIC NAME

Ceano'thus american'us L. var.
pitch' eri T. & G. is a member of
the Rhamnaceae (Buckthorn Family). The name *Ceanothus* is a
name used by Theophrastus (a
Greek who died about 285 B.C.)
for some spiny plant whose identity is unknown. The species
name *americanus* means
"of America."

DESCRIPTION

Shrubs to 1 m (40 in) tall, often
branched above, with deep, reddish roots. Leaves alternate,
broadly oblong, 5–10 cm (2–4 in)
long, lower surfaces densely hairy,
margins irregularly toothed. Flowers in elongated, branched groups
among leaves on young branches,
from May to Jul; petals 5, separate, rounded with long, narrow

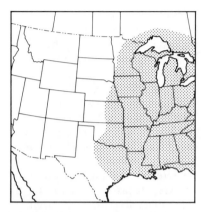

bases, white. Fruits dry, 3-lobed,
4–5 mm (± 3/16 in) wide, black,
each lobe crested; seeds dark red-
brown and glossy.

C. herbaceous var. *pubescens*
differs in having elliptic leaves,
flowers in clusters at ends of current year's branches, and fruit
lobes without crests.

HABITAT

Eastern prairies (especially rocky
hillsides), glades, and open woods.

PARTS USED

Leaves (best when dried in the
shade after being harvested from
plants in full bloom in late
spring)—fresh or dried for tea.

FOOD USE

The leaves of New Jersey tea
make a good-tasting and caffeine-
free substitute for black tea. Both
New Jersey tea species were used
by all the tribes of the Upper Missouri River region. The gnarled

roots of this shrub are often much larger than the aboveground part. On buffalo hunting trips, these large reddish-brown roots were used to make fires where timber was scarce (Gilmore, 1977, p. 50).

Indians along the Atlantic Coast (perhaps in New Jersey) probably taught the colonists the use of New Jersey tea, which was used as a patriotic substitute for black tea during the Revolutionary War. Manasseh Cutler in 1774 wrote in an account of the useful native plants found in the colonies before the Revolutionary War:

The leaves of this shrub have been much used by the common people, in some parts of the country, in room of India tea; and is, perhaps, the best substitute the country affords. They immerse the fresh leaves in a boiling decoction of the leaves and branches of the same shrub, and then dry them with a gentle heat. The tea, when the leaves are cured in this way, has an agreeable taste, and leaves a roughness on the tongue somewhat resembling that of bohea tea (a black tea from China) (Cutler, 1785, in Fernald et al., 1958, p. 271).

The English name soapbloom for *Ceanothus* and the use of the flowers of western species by Indians and pioneers as a soap substitute indicate the presence of saponin (a glucoside that makes a soapy lather and is poisonous if consumed in large quantities). Saponin is probably found in the flowers of all species (Forest Service, 1937, p. B40). Perhaps the preparation technique mentioned by Cutler was to remove traces of this soapy substance from the leaves. However, tea made from dried leaves does not taste soapy and the amounts of saponin in it must be insignificant. It is best to pick leaves while the plant is in full bloom. These should be dried in a shady or dark and dry place.

CULTIVATION

New Jersey tea is a pretty shrub that can be cultivated for its fragrant white flower clusters and as a tea plant. Julian Steyermark (1981, p. 1030) remarks that this plant "is sometimes grown as an ornamental, and its use for such should be more extensively recommended." There are many named ornamental varieties of *Ceanothus*. They will grow well in a light, well-drained soil in a sunny location. New Jersey tea can be propagated by root division in the fall, softwood cuttings forced in a greenhouse in the spring, or by seed. Propagation by seed appears to be the most successful. Seed should be planted immediately in the spring after scarification and soaking in hot water for 30 minutes (Smith and Smith, 1980, p. 85).

Chenopodium berlandieri
Lamb's Quarters

C. Kuhn
©86

Lamb's quarters, lambsquarter, goosefoot, pitseed goosefoot, wild spinach, Indian spinach, pigweed, frostblite, fat-hen, bacon weed, poulette, and chou grass.

INDIAN NAMES

The Pawnee name is "kitsarius" (green juice), referring to use of the plant for a green dye (Gilmore, 1977, p. 26). The Dakota called lamb's quarters "wahpe toto" (greens) (ibid.). The Lakota name is "canxlogan inkpa gmigmela" (small end [of the leaves?] rounded weed) (Munson, 1981, p. 233).

SCIENTIFIC NAME

Chenopod'ium berlandier'i Moq. is a member of the Chenopodi-aceae (Goosefoot Family). *Chenopodium* is Greek for goose-foot, in reference to the shape of the leaves of some species. The species name *berlandieri* is in honor of the plant's scientific discoverer, Jean Louis Berlandier (1805–1851), a French physician who collected plants in Texas and northern Mexico.

DESCRIPTION

Annual herbs with slender, arching branches, usually less than 1 m (40 in) tall. Leaves alternate, 4-angled to egg-shaped or lance-shaped, 2–4 cm (¾–1⅝ in) long, thick, surfaces mealy, margins irregularly wavy, sometimes un-

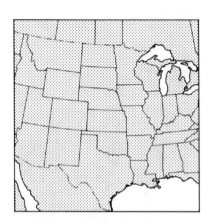

pleasantly scented. Flowers small, in clusters grouped into dense heads or branches at tops of plants, from Jul to Sep; sepals 5, each with a sharp ridge, greenish, no petals. Fruits dry, inflated, 1.2–1.5 mm (±¹⁄₁₆ in) in diam; seeds with honey-combed covering.

HABITAT

Roadsides, pastures, and waste ground.

PARTS USED

Young leaves (spring and early summer)—raw or cooked; seeds (fall, early winter)—dried, then cooked or ground into meal.

FOOD USE

Lamb's quarters is an ancient food plant of the prairies that was gathered wild, and perhaps cultivated, by the Indians for its tender spring greens and fall seeds. There has been some confusion over which species were native and which were actually cultivated. The various species are similar in their appearance and use.

The Dakota and Omaha used lamb's quarters for food and cooked the young and tender greens. Melvin Gilmore, who studied the ethnobotany of both these tribes, reported (1977, p. 26) that the use of *Chenopodium album*, L. "appears to be so long established that the fact of its introduction seems now unknown to the Indians." It is unclear whether *C. album* is introduced or was actually native to some parts of our country. The scientific name is now best restricted to the weedy, Old World phase, but it has been used for the entire lamb's quarters complex. It is known that *C. berlandieri* and several other species are native and have been used since prehistoric times as food.

The Kiowa were reported to have used lamb's quarters for food (Vestal and Schultes, 1939, p. 25). The Pawnee ate lamb's quarters, and it was also used as a green dye for bows and arrows (Gilmore, 1977, p. 26).

Numerous reports of the use of lamb's quarters from the Southwest indicate its importance as a food source and suggest more widespread use in the Prairie Bioregion. *C. leptophyllum* Nutt. ex Moq. was called "small seeds" by the Zuni, who stated that the seeds of this plant, along with those of the sage, *Artemisia*

wrightii, were among their principal foods when they first reached this world (Stevenson, 1915, p. 66). It was used in the following manner: "The seeds are ground, mixed with corn meal seasoned with salt, and made into a stiff batter, which is formed into balls or pats and steamed. Upon first reaching this world the seeds were prepared without the meal, as the Zuni had no corn at that time. Now the young plants are boiled either with or without meat, and are greatly relished" (ibid.).

Edward Palmer reported in "Food Products of the North American Indians" (1871, p. 419):

The young tender plants are collected by the Navajoes, the Pueblo Indians of New Mexico, all the tribes of Arizona, the Diggers of California, and the Utahs, and boiled as herbs alone, or with other food. Large quantities also are eaten in the raw state. The seeds of this plant are gathered by many tribes, ground into flour after drying, and made into bread or mush. They are very small, of a gray color, and not unpleasant when eaten raw. The peculiar color of the flour imparts to the bread a very dirty look, and when baked in ashes it is not improved in appearance. It resembles buckwheat in color and taste, and is regarded as equally nutritious. The plant abounds in the Navajo country.

Lamb's quarters has very moist

leaves and the Hopi packed them around other foods, such as yucca shoots, when these were pit-baked together (Niethammer, 1974, p. 112).

Lamb's quarters was mentioned by the early white explorers and travelers of the Prairie Bioregion. It is a "weedy" species and commonly found in disturbed habitats. In 1845, when Lieutenant James Abert was traveling on the Sante Fe trail, he often reported seeing the lamb's quarters plant along the trail. Specifically, on July 9 (in central Kansas), he reported: "We collected some lamb's quarter and had it cooked" (McKelvey, 1955, p. 985).

Lamb's quarters greens are quite nutritious and tasty. They are my favorite of all the wild edibles I harvest from my vegetable garden. Either raw in a salad, or steamed and served with butter, they are a treat. The seeds are reported to be "of good flavor and highly nutritious, tasting somewhat like buckwheat but with the characteristic 'mousey' flavor distinctive of this group of plants" (Fernald et al., 1958, p. 179). Lamb's quarters seeds are about equal to corn in the number of calories they contain, but have significantly more protein and fat (Asch and Asch, 1978, p. 307). The cooked greens contain more than three times as much calcium as cooked spinach and also have more vitamin A and C (Watt and Merrill, 1963, pp. 37, 59).

CULTIVATION

Seeds of lamb's quarters are commonly found in excavations of prehistoric sites. One plant can produce up to 100,000 seeds. Since lamb's quarters thrives in ecologically disturbed habitats, such as former Indian village sites, it has been suggested that the seeds could possibly get mixed naturally into the archaeological remains at these sites or be carried there by rodents. However, in eastern North America, there are at least eight instances of *Chenopodium* seeds being recovered from storage pits and human coprolites (Asch and Asch, 1977, p. 6).

Some of the oldest lamb's quarters seeds, *C. berlandieri*, are from the Koster site in southwest Illinois, which was occupied from 6500 to 3000 B.C. (Asch and Asch, 1982, p. 9; Asch and Asch, 1985, p. 174). Melvin Gilmore (1931b, p. 97) found seeds of lamb's quarters in the ruins of the Ozark Bluff-dwellers, indicating that the plants were cultivated: "Sheaves of seed heads of a species of *Chenopodium* were found put away with other stored seeds, probably indicating the cultivation of this plant. Species of *Chenopodium* were cultivated for food in South America, Central America and Mexico in pre-Columbian times, and still are at the present time."

Hugh Wilson, a chenopod specialist, has recently made the case that seeds from the Bluff-dwellers were of Mexican origin, derived from *C. berlandieri* subspecies *nuttalliae*, which is still grown in the central Mexican highlands (Asch and Asch, 1982, p. 9). In that area, lamb's quarters is "double-harvested, that is, the greens are gathered early in the growing season and then the seeds are collected later when they ripen" (Ford, 1981, p. 20). This pattern of use is undoubtedly prehistoric. Greens and dried leaves do not preserve well at archaeological sites, but the presence of lamb's quarters seeds suggests that the leaves also may have served as sources of food.

In our region, over 4,500 lamb's quarters seeds have been found at the Coffey site (located near Tuttle Creek Reservoir in northeast Kansas), which was occupied from approximately 3730 to 3130 B.C. (Schmidts, 1978, 1981, in Adair, 1984, p. 40). Lamb's quarters seeds were found in such great quantity and frequency at the Mitchell site (near Mitchell, South Dakota), that "there can be no doubt that they were intensively collected" (Benn, 1974, p. 229). This site was occupied from 1250 to 1450 A.D. and it is quite possible that Indians actually semicultivated this plant or at least encouraged it.

J. W. Blankenship, botanist for the Montana Agricultural College Experiment Station, indicated that Indians scattered the seeds of lamb's quarters and wrote (1905, p. 9) that the seeds "were gathered by the Indians for food, which may account in part for its wide distribution and abundance in the state as is probably the case with the sunflower (*Helianthus annuus* L.) and the horseweed (*Iva xanthifolia* Nutt.). The seeds were ground into flour and made into bread. The young plants are also used by Indians and whites as a pot-herb."

At the Ross site, along the Oldman River (northeast of Coaldale, Alberta), a cache of lamb's quarters seeds was found among other cultural debris that identified the site as being occupied from 1500 to 1600 A.D. by the Blackfoot (Johnston, 1962, p. 129). While it was not suggested that they were cultivating the plant, the Blackfoot did collect it and may have encouraged it at these collection places.

Lamb's quarters grows very easily from seed. It does not need orderly cultivation. It can be encouraged to grow in some selected place or in a part of the garden that is seldom used. It likes and does best in very rich ground, such as an old manure or hay pile, which has been tilled in the early spring or fall.

Cirsium undulatum
Wavy-leafed Thistle

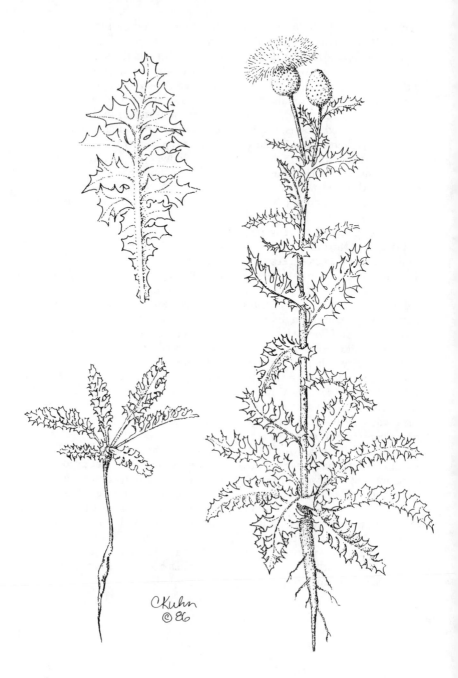

CKuhn
© 86

Wavy-leafed thistle and thistle.

INDIAN NAMES

The Kiowa name for the yellow-spined thistle, *Cirsium ochrocentrum* Gray, is "sengts-on" (thistle) (Vestal and Schultes, 1939, p. 59).

SCIENTIFIC NAME

Cir'sium undula'tum (Nutt.) Spreng. is a member of the Asteraceae (Sunflower Family). *Cirsium* is from the Greek "kirsion," the ancient name of a kind of thistle. The species name *undulatum* means "wavy," referring to the leaf margins.

DESCRIPTION

Perennial herbs with erect stems 4–10 dm (16–40 in) tall, densely covered with white hairs. Leaves alternate, elliptic, those at base 10–30 cm (4–12 in) long, lobed and wavy, lobes tipped with yellow spines; upper leaves oval to lance-shaped, shallowly lobed or nearly entire, becoming smaller near tops of plants. Flower heads globular, with cuplike bases of spiny modified leaves, from Jun to Aug; florets tubular, purple. Fruits dry, small, each topped by a prominent tuft of white, stiff hairs.

HABITAT

Prairies and waste ground.

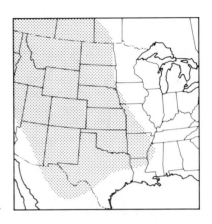

PARTS USED

Leaves (spring)—trimmed and cooked; stalks (May, June)—peeled, raw; flowers and seeds (late summer, fall)—raw or cooked; roots (late fall, spring, or summer)—raw or cooked.

FOOD USE

This spiny plant can become a serious weed in heavily grazed pastures. Except for its spines, which cover the stems, leaf edges, and the base of flowers, most of the plant is edible, and it has been widely used as a food source. Elias Yanovsky lists eight species of *Cirsium* in his "Food Plants of the North American Indians" (1936, pp. 60–61).

The Comanche were reported to eat the roots of the wavy-leafed thistle (Carlson and Jones, 1939, p. 521). The older members of the Kiowa tribe in 1939 remembered when the roots of the yellow-spined thistle (*C. ochrocentrum*) were used as food (Vestal and Schultes, 1939, p. 59). Among the Chiricahua and Mescalero Apache, the seeds of *C. pallidum* were boiled and eaten in the same manner as sunflower seeds, or they were ground into flour, and the dough was baked (Opler, 1936, p. 49).

The Blackfoot would dissect the head of thistles and eat the flower stalk fresh (Hellson, 1974, p. 102). The Cheyenne traveled to the Bighorn Mountains to obtain the stems of the thistle, *C. edule,* which they spoke of as a fruit and considered to be a great luxury (Grinnell, 1962, p. 191). The Gosiute of the Great Basin also ate the stems of the wavy-leafed thistle (Chamberlin, 1911, p. 366).

Thistles were not generally a major food source, but they were used when needed. The roots of the thistle, *C. foliosum,* more than any other food, kept Truman Everts from starving when he was lost for more than a month in the Yellowstone Park region in 1870 (Medsger, 1966, p. 201).

Thistles are literally a pain to collect. When the stems are gathered, gloves should be worn and the spiny leaves stripped off in the field with a long-handled knife. The spines are so sharp that sometimes they will even poke through leather gloves! The raw, peeled stems have a taste resembling artichokes and their texture is similar to that of celery. When cooked, they are tender and flavorful. The

leaves can also be used as a pot herb. Scissors can be used to cut off the spines before the leaves are cooked.

At the Kansas Area Watershed (KAW) Council winter camp on December 22, 1984, at Kanopolis Reservoir in central Kansas, I harvested some wavy-leafed thistle roots. Very few edible plants are available in the prairies in winter. These roots were easy to identify and locate from their winter rosettes (small circular clusters of silvery blue-green leaves near the ground). After washing the top of the root and crown, I boiled them for five minutes and found them to be tender and tasty.

The musk thistle, *Carduus nutans* L., is a noxious weed in some states and it is illegal to grow it. Other thistles are often sprayed with herbicides to kill them, so to avoid ingesting herbicides, you should be careful where thistles are harvested. Any thistles or other plants with bizarrely shaped parts or burned places on their leaves have probably been sprayed.

CULTIVATION

While I was growing up on a farm in southern Nebraska, I learned to fight thistles with a hoe, shovel, and corn knife as invasive weeds in our pastures. Therefore, the thought of cultivating them is not too exciting to me. However, someone might be interested in growing them for their attractive flowers or food uses. They can be propagated by seed, or winter rosettes with a good piece of the taproot can be transplanted.

Thistles spread easily, because their seeds are dispersed by the wind. If grown in a garden, their flower heads should be cut off before they mature, so that seeds will not float to your neighbor's land.

Claytonia virginica
Spring Beauty

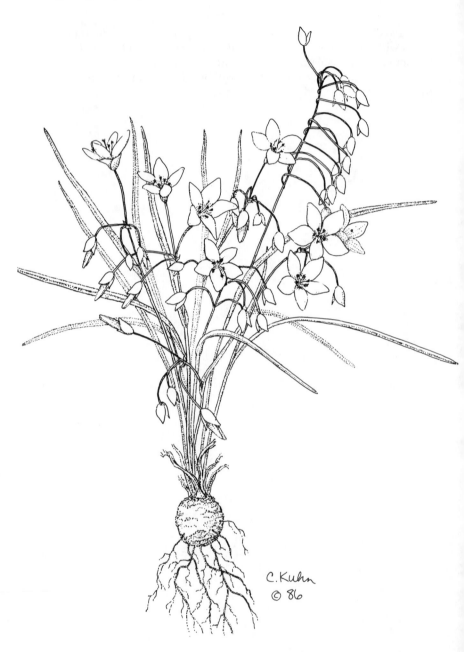

C. Kuhn
© 86

Spring beauty, Virginia spring-
beauty, and fairy spuds.

INDIAN NAMES

None were found in the sources
consulted.

SCIENTIFIC NAME

Clayton'ia virgin'ica L. is a member of the Portulacaceae (Purslane
Family). *Claytonia* is named in
honor of Dr. John Clayton, botanist and notable plant collector of
Colonial days. The species name
virginica means "of Virginia."

DESCRIPTION

Perennial herbs from round
corms, stems single, fleshy, 1–3
dm (4–12 in) tall. Leaves opposite,
narrowly lance–shaped, 5–20 cm
(2–8 in) long, fleshy. Flowers in
elongated groups at tops of stems,
from Feb to Jul; petals 5, separate,
oval, 9–14 mm (⅜–½ in) long,
white or rose with pink or purple
veins. Fruits dry, roundish, 4 mm
long, edge of segments rolling in-
ward as they open; seeds round,
shiny, blackish-brown.

HABITAT

Open woods, prairies, meadows,
and rocky ledges.

PARTS USED

Young leaves and plants (spring)—
raw or cooked; roots (late spring

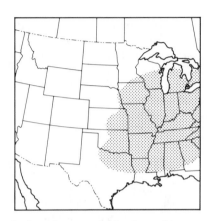

before the tops die back, making
it difficult to find the roots)—raw
or cooked.

FOOD USE

This early spring wildflower lives
up to its name: the petals of its
small pinkish-white flowers are
highlighted with a brilliant pink
to purple veination pattern. The
entire plant is edible raw or
cooked, although the half-inch diameter corms are the part usually
mentioned as edible. When eaten
raw, they taste starchy and a little
sweet. However, the beauty of this
plant is greater than the nourish-
ment it provides, so it should not
normally be uprooted.

There are no specific references
to spring beauty as a food source
for Indians who lived in the Prairie Bioregion. Dr. V. Havard (1895,
p. 107) mentioned it as being
found in the eastern states and
having deep edible bulbs "whose
crisp flesh and nutty flavor were
much prized by the natives."

Thomas Nuttall, a botanist and an early plant explorer of the west, was one of the first to describe it in a prairie habitat. He found spring beauty in full bloom in some prairie openings and meadows northwest of what is today Little Rock, Arkansas (Nuttall, 1980, p. 87).

Spring beauty is not found very far west or north in the prairies. Dr. F. V. Hayden, in *Botany Report of the Secretary of War* (1859, p. 731), reported its presence in rocky woods along the Missouri River as far north as Council Bluffs. The spring beauty is primarily a woodland wildflower, but occasionally grows in meadows and prairies. In *Edible Wild Plants of Eastern North America*, it is mentioned as abundant in some meadows: "The succulent, opposite-leaved young plants, which often abound in spring in rich woods and open glades, are a possible pot herb. Only in regions where the plants are superabundant, however, would the quantity be sufficient to repay digging for the deeply buried roots. In some regions the plants cover many acres of wooded slopes or meadows . . ." (Fernald et al., 1958, p. 197). On my walk across the Kansas and Colorado prairie in 1983, I encountered spring beauty in bloom in both a woodland park in Kansas City, Kansas, and at the western edge of its range in a private pasture in Ellsworth County, in central Kansas.

CULTIVATION

Hortus Third (Bailey, 1976, p. 280) notes that spring beauty can be transplanted from the wild to moist, shady locations and rock gardens. It can also be grown in a sunny location. Spring beauty should not be transplanted unless a large wild population is found. It sometimes spreads to lawns and parks without cultivation.

Cleome serrulata
Rocky Mountain Bee Plant

C.Kuhn
©86

Rocky Mountain bee plant, bee spiderflower, cleome, spider-flower, stinkweed, stinking clover, and guaco.

INDIAN NAMES

The Lakota name is "wahpe'-h'eh'e" (ragged leaf) (Rogers, 1980a, p. 61). In the Southwest, the Zuni call it "a'pilalu" (hand, many seeds) referring to the handlike shape of the leaf (Stevenson, 1915, p. 69).

SCIENTIFIC NAME

Cleo'me serrula'ta Pursh is a member of the Capparaceae (Caper Family). Octavius Horatius, a Roman physician of the fourth century, used the name *Cleome* for a related plant in the Mustard Family and Linneaus later transferred this name to its present use. The species name *serrulata* means "little saw" in reference to the sawtooth leaf margins on some individual plants.

DESCRIPTION

Annual herbs, 2–15 dm (8–60 in) tall, branched. Leaves divided into 3 fingerlike segments, narrow, 2–6 cm (.8–5 in) long. Flowers in dense, elongated groups at ends of branches, from Jun to Aug; petals 4, separate, 8–12 mm (⁵/₁₆–½ in) long; oblong, bright pink to purplish; stamens 6, much longer than petals. Fruits dry, narrow to

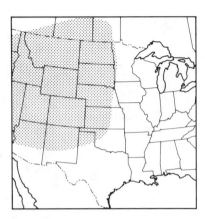

spindle-shaped, sharply pointed at both ends, 2–8 cm (¾–3¼ in) long; seeds egg-shaped, blackish-brown mottled.

HABITAT

Sandy or rocky prairies and waste places.

PARTS USED

Leaves (late spring and summer)—cooked; seeds, (late summer, fall)—ground for flour.

FOOD USE

It may seem surprising that this attractive, strong smelling, and hairy plant was used as a food source. Its leaves were cooked for greens, like spinach, although it was necessary to boil them for a long time to remove the alkali taste.

Accounts from the Southwest tell of its use by the Indians there; it may have been used similarly by Indians of the Prairie Bio-region. Edward Castetter (1935, p. 24) an anthropologist, reported that Rocky Mountain bee plant was "one of the most important native plants in use by the Pueblo Indians of New Mexico, having outstanding significance as a food plant even today, although this was obviously more marked before the introduction of some of the cultivated plants." The Zuni boiled the tender leaves with corn (on or off the cob) and highly seasoned the concoction with chili. Large quantities of the leaves were also gathered and hung indoors to dry for winter use (Stevenson, 1915, p. 69). Among the Tewa of Hano, the plant was of sufficient economic importance to be mentioned in songs, along with the three plants most cultivated—corn, pumpkins, and cotton (Niethammer, 1974, p. 104).

The Tewa used this plant, which they called guaco, for paint and for food:

This is a very important plant with the Tewa, inasmuch as black paint for pottery decoration is made from it. Large quantities of young plants are collected, usually in July. The plants are boiled well in water; the woody parts are then removed and the decoction is again allowed to boil until it becomes thick and attains a black color. This thick fluid is poured on a board to dry and soon becomes hardened. It may be kept in hard cakes for an indefinite period. When needed these are soaked in hot water until of the consistency needed for paint.

Guaco is also used as a food. The hardened cakes are soaked in hot water, and then fried in grease (Robbins et al., 1916, pp. 58–59).

The Hopi gathered the young Rocky Mountain bee plant in spring, and served it along with cornmeal porridge and a small quantity of salt (ibid.). It has been reported that the flowers were also highly esteemed (Fewkes, 1896, p. 16) and that the basal leaves were eaten with green corn (Hough, 1897, p. 37).

The Navaho gathered the greens for food and to eat during the Nightway ceremony (Wyman and Harris, 1951, p. 25). They also gathered the seeds and ground them into a meal for gruel or bread (Harrington, 1967, p. 72).

CULTIVATION

Richard Yarnell, an anthropologist, has recently reported archaeological evidence suggesting that *Cleome* was collected in quantity by prehistoric people of the Southwest, and that it may have been encouraged through tending or even planted (Ford, 1981a, p. 21). Rocky Mountain bee plant, as its name suggests, attracts bees because of its nectar. It has sometimes been cultivated for this purpose. It can also be planted for its attractive flowers and a few greens can be eaten from the plants early in the spring without slowing the growth of the plants significantly. A species native to southeast Brazil and Argentina, *C. hassleriana* Chodat., is frequently cultivated in the flower garden. Its cultivars include "Great Pink," "Pink Queen," "Rosea," "Alba," and "Snow Crown." Both our native species and this one are annuals. Their seeds can be planted in the spring after danger of frost is past.

Comandra umbellata
Bastard Toadflax

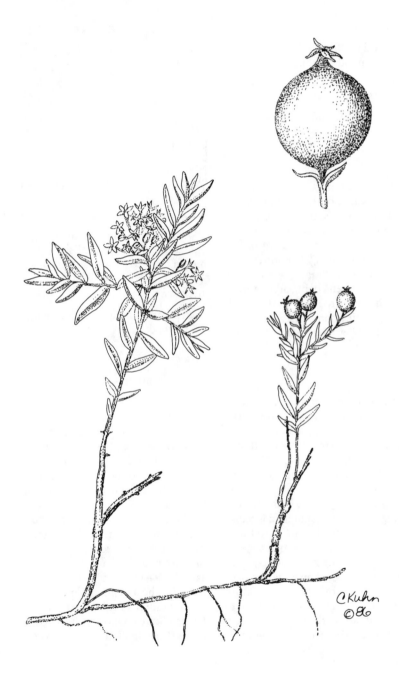

CKuhn
©86

COMMON NAMES

Bastard toadflax, false toadflax, and comandra.

INDIAN NAMES

None were found in the sources consulted.

SCIENTIFIC NAME

Coman'dra umbella'ta (L.) Nutt. is a member of the Santalaceae (Sandlewood Family). *Comandra* comes from the Greek word "kome," which means "tufts of hairs" and "aner," "andros," which means "man" in reference to the stamens being adherent to the hairy-tufted base of the sepals. The species name *umbellata* refers to the arrangement of flowers in umbels.

DESCRIPTION

Perennial herbs from extensive, horizontal rhizomes, stems clustered, 0.7–5 dm (3–20 in) tall. Leaves alternate, linear to elliptic or egg-shaped, 0.7–4 cm (¼–1⅝ in) long, some thin and green, others thick and gray-green. Flowers in small, flat-topped clusters arranged in loose, branched groups, at end of stems from Apr to Jul; sepals fused into greenish tube, then separating into 5 lobes, pointed to egg-shaped, white or sometimes pink, no petals. Fruits fleshy, roundish, 4–7 mm (³⁄₁₆–¼ in) in diam, with tubes and lobes of flowers sometimes persisting

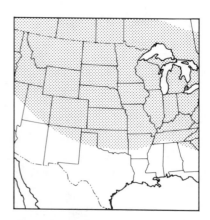

above, green then maturing to chestnut brown or purplish brown.

HABITAT

Prairies and infrequently in rocky, open woods.

PARTS USED

Fruits (summer)—raw.

FOOD USE

Bastard toadflax is a nondescript, parasitic prairie plant. Although its use for food by Indians of the prairies has not been reported, it was probably an occasional, minor food source.

The edible, urn-shaped fruits of bastard toadflax are seldom found in quantity and rarely provide more than a pleasant-tasting sample. Bastard toadflax is reported to be a popular plant among Indians in the western United States, who ate the sweet fruits (Fernald et al., 1958, p. 166).

In *Wild Edible Plants of the Western United States* (1970, p. 115), Donald Kirk states that the fruits, eaten raw, are best when slightly green, although they are still quite edible when fully mature and a brown color. I have tasted the green fruits during August in Kansas and found them to be bland—palatable, but with no interesting flavor to recommend their consumption.

Edward Palmer in his 1878 report suggests that one should not eat quantities of bastard toadflax: "This plant yields a small nut which is eaten raw by the Pah-Utes and the white children of Utah. If eaten too freely it produces nausea."

Bastard toadflax is often overlooked in the prairie because of its small size and light green color. It parasitizes the roots of other plants, but is more properly called a hemi-parasite because it has its own green leaves that photosynthesize and provide energy for growth. It is similar to mistletoe, robbing its hosts of little more than water (Stevens, 1961, p. 335).

CULTIVATION

Bastard toadflax is sometimes transplanted to wildflower gardens for its attractive small white flowers, which bloom for a long time in the spring. In the *Prairie Propagation Handbook*, it is suggested that, since it is parasitic, it should be planted near other plants (Rock, 1977, p. 28).

Corylus americana
Hazelnut

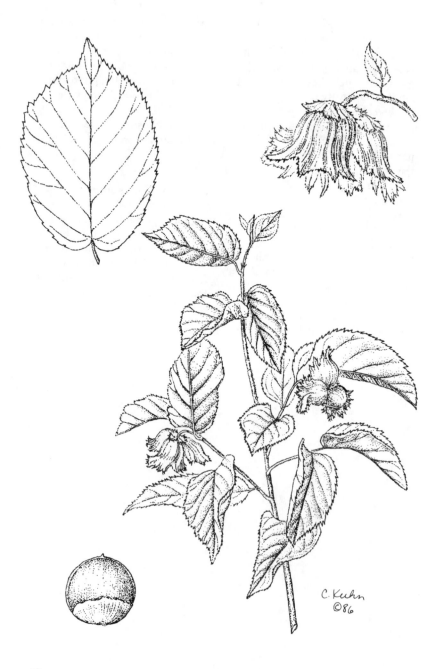

C. Kuhn
©86

Hazelnut, American hazelnut, hazel, filbert, and American filbert.

INDIAN NAMES

The Dakota name is "uma" and the name "uma-hu" means "hazel bush"; the Omaha and Ponca name is "unzhinga" and "unzhinga-hi" meaning "hazel bush," and the Winnebago name is "huksik," not translated (Gilmore, 1977, p. 22).

SCIENTIFIC NAME

Cor'ylus american'a Walt. is a member of the Betulaceae (Birch Family). *Corylus* is the Latin name for "hazelnut." The species name *americana* means "American."

DESCRIPTION

Shrubs to 3 m (10 ft.) tall, young twigs hairy. Leaves alternate, oval, 1–12 cm (⅜–4¾ in) long, rounded to heart–shaped at bases, lower surfaces hairy, margins finely toothed. Flowers in long, hanging clusters, male and female separate, appearing before leaves from Mar to May; each one small, greenish, no petals. Fruits dry, hard, 1–1.5 cm (⅜–⅝ in) in diam, solitary or clustered, enclosed in a ragged-edged husk, ripening in Sep.

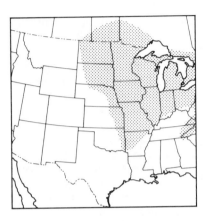

HABITAT

Dry or moist thickets and woodland borders, in valleys and uplands.

PARTS USED

Nuts (late summer or fall)—eaten fresh, ground into meal, or cooked.

FOOD USE

The distribution of hazelnuts extends into the prairie from the eastern woodlands. These tasty nuts are hard to find on prairie/woodland borders because they are a favorite food of birds and squirrels, and additionally, the bushes seem to be less productive in the western portion of their range, perhaps because of the drier conditions. The nuts were a food source of the Omaha, Ponca, Winnebago, Dakota, and probably other tribes of the eastern portion of the Prairie Bioregion (Gilmore,

1977, p. 22). They ate the nuts raw (sometimes with wild honey) or used them as the basis for soup (ibid.). The Potawatomi, who previously lived in the Great Lakes Bioregion, were reported to be especially fond of the nuts late in the summer when they were in the milk stage and had not yet hardened (Smith, 1933, p. 97). It is quite likely that other tribes also used them at this stage, because the nut meats are soft and sweeter than when fully mature.

Hazelnuts have been found in archaeological remains along the eastern border of the Prairie Bioregion, such as the Middle Woodland sites in Illinois. Nancy and David Asch (1980, p. 155) report that the nuts were dominant at Dickson Camp and Pond sites near the Illinois River in southwestern Illinois. Fragments of hazelnut shells were also found at the Archaic Nebo Hill site, near Kansas City, Missouri, which was probably occupied sometime between 1600 and 270 B.C. (Adair, 1984, p. 40).

The following historical accounts further document the occurrence of hazelnuts at the woody margins of prairie. The boundary between the prairie and the eastern deciduous forest is not a distinct line. The two types of vegetation intermingle and their border has fluctuated east and west during regional climatic changes that have occurred over the years.

Captain William Clark, of the Lewis and Clark expedition, reported this observation of hazelnuts and prairie on August 1, 1804, along the Missouri River, north of what is today Omaha, Nebraska: "The Prarie which is situated below our Camp is above the high water leavel and rich covered with Grass from 5 to 8 feet high interspersed with copse of Hazel, Plumbs, Currents (like those of the U.S.) Rasberries & Grapes of Dift. Kinds. also producing a variety of Plants and flowers not common in the United States" (Thwaites, 1904, 1: 96).

Edmund Flagg in 1836 observed how the landscape immediately west of St. Louis had changed from a "shrubless waste" (prairie), over the last 30 years:

The prospect in this direction is boundless for miles around, till the tree-tops blend with the western horizon. The face of the country is neither uniform or broken, but undulates almost imperceptibly with thickets of wild-plum, the crab-apple, and the hazel. Thirty years ago, and this broad plain was a treeless, shrubless waste, without a solitary farmhouse to break the monotony. But the annual fires were stopped; a young forest sprang into existence; and delightful villas and country seats are now gleaming from the dark foliage in all directions (Thwaites, 1906, 26: 162).

Hazelnuts are very tasty and the wild ones seem to have even more flavor than the domesticated European cultivars. Hazelnuts, like other nuts, are a good energy source and are high in calories and protein (Watt and Merrill, 1963, p. 88). They are relished by squirrels and bluejays and other birds.

CULTIVATION

The hazelnut should be cultivated for its tasty nuts and the attractively colored fall foliage. The nuts sold in grocery stores are a cross between the European hazelnut and the giant hazelnut. These require a special growing climate, such as that of the Willamette Valley of Oregon and around the Black Sea in Turkey. There are some cultivars, such as "Bixby," "Buchanan," "Potomac," and "Reed," that are crosses between the European hazelnut and our American hazelnut and are more adapted to eastern North America. Hazelnuts thrive in a deep, friable, well-drained soil and can be planted in the same manner as other shrubs and nursery stock. In the home garden, at least two different cultivars should be planted a few yards apart to ensure cross-pollination (Bailey, 1976, p. 480).

Coryphantha vivipara
Pincushion Cactus

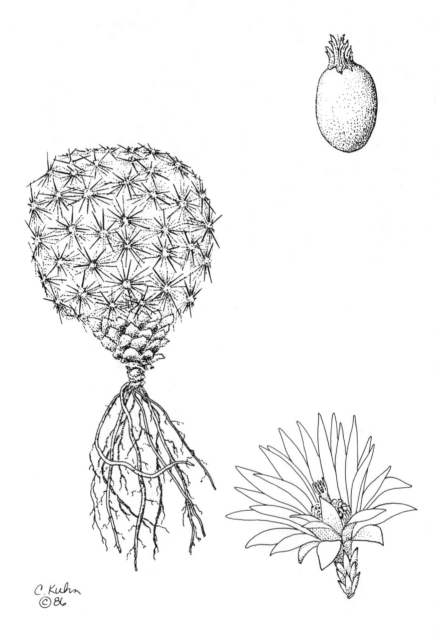

C. Kuhn
©86

COMMON NAMES

Pincushion cactus, nipple cactus, sprouting pincushion cactus, cushion cactus, mammillaria, purple mammillaria, bunch cactus, ball cactus, and golfball cactus.

INDIAN NAMES

The Blackfoot name is "ost-staxie-mon" (Johnston, 1970, p. 316). One translation is "wild figs" (Hellson, 1974, p. 103), probably in reference to the fruit. The Cheyenne names are "matahesono" and "mataha" (no translations given) (Hart, 1981, p. 16).

SCIENTIFIC NAME

Coryphan'tha vivi'para (Nutt.) Britt. & Rose is a member of the Cactaceae (Cactus Family). *Coryphantha* means "head, top, flower" because the flowers develop near the apex of the stem. The species name *vivapara* means "bringing forth its young alive," referring to the way this species vegetatively produces small pincushion-shaped offspring, clustered around its base.

DESCRIPTION

Perennials with 1 to several stems, globose to cylindrical, to 7 cm (3 in) tall, covered with spirally arranged, knoblike projections, bearing clusters of spines. No leaves. Flowers solitary near tops of stems from May to Aug; 2.5–4 cm (1–1½ in) long, with

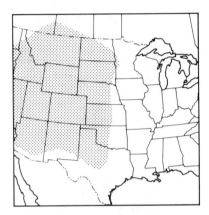

many segments, pink to reddish purple. Fruits fleshy, oblong to club-shaped, 1–2.5 cm (⅜–1 in) long, green; seeds brown, pitted.

HABITAT

Dry rocky prairie hillsides and uplands, frequently growing on limestone or sandy soils.

PARTS USED

Fruit (late summer)—raw; flowers (summer)—raw; entire plant (despined)—eaten raw or cooked.

FOOD USE

Buffalobird Woman reported during the 1910s that Hidatsa Indian girls would collect a few fruit from the pincushion cactus late in the summer and eat them fresh (Nickel, 1974, p. 67). Wolf Chief, also an Hidatsa, reported that a traveler who needed food could eat this cactus after it had been roasted (a process that would burn off the spines) (ibid.). The fruits

103

were also eaten by the Blackfoot as a confection (Hellson, 1974, p. 103). The Cheyenne cooked them by boiling, either freshly picked or after they had been dried (Hart, 1981, p. 16).

The Crow were reported to have eaten the red, ripe fruit of the closely related *C. missouriensis* (Sweet) Britt. & Rose (Blankenship, 1905, p. 15). Dr. V. Havard of the U.S. Army reported in "Food Plants of the North American Indians" (1895, p. 116) that he had eaten the red berries of pincushion cactus from the Upper Missouri with "great relish." I have found the red fruits from the previous summer still clinging between the spines of this cactus in late April on a Flint Hills pasture in Kansas. The fruits were still moist and slightly sweet, so that they had been available as a food source all winter.

The flowers of pincushion cactus also were a food source for prehistoric Indians who lived in southwest Texas from 800 B.C. to 500 A.D. Studies of the coprolites from the Amisted area revealed pincushion cactus pollen, which indicates that the flowers were being eaten (Bryant, 1974, p. 407).

CULTIVATION

Pincushion cactus is a slow-growing plant that can be cultivated for both its showy pink flowers and its fruits. It can be successfully transplanted to a sunny, well-drained location.

Cucurbita foetidissima
Buffalo Gourd

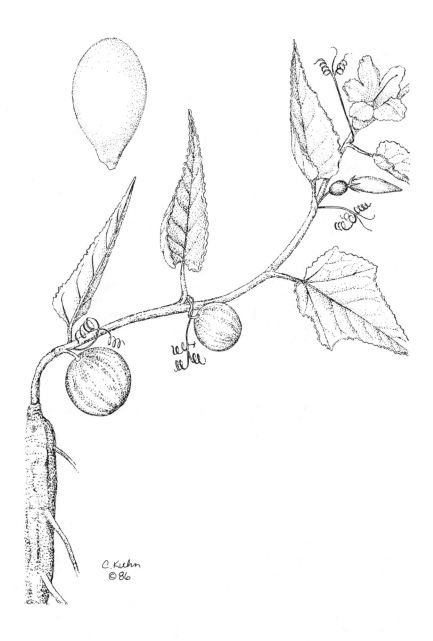

C. Kuhn
© 86

Buffalo gourd, Missouri gourd, coyote gourd, fetid gourd, fetid wild pumpkin, wild pumpkin, chili coyote, and calabezella.

INDIAN NAMES

The Osage call it "monkon tonga" (big medicine) and "monkon nikasinga" (human being medicine) (LaFlesche, 1932, p. 100). The Dakota call it "wagamun pezhuta" (pumpkin medicine) (Gilmore, 1977, p. 64). The Omaha and Ponca call it "niashiga makan" (human being medicine), and they also distinguished between plant forms that they identified as male and female (ibid.). The above names are all in reference to the mystical and medicinal powers of the plant.

SCIENTIFIC NAME

Cucur'bita foetidis'sima H.B.K. is a member of the Cucurbitaceae (Cucumber Family). *Cucurbita* means "gourd," and the name *foetidissima* means "ill-smelling," because the leaves have a very rank odor.

DESCRIPTION

Perennial herbs, trailing on ground, stems often several meters (yards) long. Leaves alternate, triangular to egg-shaped, 1–2 dm (4–8 in) long, upper surfaces rough, lower surfaces hairy, margins irregularly toothed, with unpleasant smell. Flowers solitary

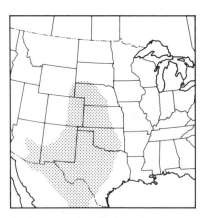

among leaves from Jun to Aug; petals 5–12 cm (2–5 in) long, united in tube below, separating into 5 lobes above, yellow. Fruits fleshy, rounded, 5–10 cm (2–4 in) across, greenish orange; seeds egg-shaped, flattened and cream colored.

HABITAT

Prairies (in dry soil), railroad track right-of-ways, and waste ground.

PARTS USED

Seeds (late summer, fall)—ground for mush; roots (fall–winter)—processed for their starch and protein.

FOOD USE

With the exception of the flesh inside the seed coat, all parts of buffalo gourd plants are extremely bitter and must be processed before they can be used as food. Buffalo gourd was not reported as a native food source in the Prairie

Bioregion. However, the seeds were mentioned as a food of the Tewa of New Mexico (Vestal and Schultes, 1939, p. 54) and Indians of Arizona and Southern California (Palmer, 1878, p. 651), who ground them finely and made a mush that was eaten with enthusiasm.

The Omaha, Ponca, Dakota and other tribes used the root of the buffalo gourd medicinally. When they dug the enormous ginseng-shaped root they took special precautions, because they thought that injury to the root would result in injury to themselves or their relatives. Those considering harvesting the roots for propagation or experimentation may want to remember this. Melvin Gilmore, an ethnobotanist, reported in 1919 (1977, p. 64):

This is one of the plants considered to possess special mystic properties. People were afraid to dig it or handle it unauthorized. The properly constituted authorities might dig it, being careful to make the prescribed offering of tobacco to the spirit of the plant, accompanied by the proper prayers, and using extreme care not to wound the root in removing it from the earth. A man of my acquaintance in the Omaha tribe essayed to take up a root of this plant and in doing so cut the side of the root. Not long afterward one of his children fell, injuring its side so that death ensued, which was ascribed by the tribe

to the wounding of the root by the father.

Gilmore also reported that "when I have exhibited specimens of the root in seeking information, the Indians have asked for it. While they fear to dig it themselves, after I have assumed the risk of so doing they are willing to profit by my temerity; or it may be that the white man is not held to account by the Higher Powers of the Indian's world" (ibid.).

Harvesting the root of the buffalo gourd is no small task. Some roots are as big as a person; one extremely large root was found that weighed 178 pounds. The amazing ability of this plant to photosynthesize can be illustrated by the description of this root and plant:

Growing from the broad apex of this root, which had a maximum circumference of 4.7 feet, were 60 short, vertical, perennial stems which produced a total of 360 annual shoots. These shoots spread out on all sides of the central crown covering an area of 40 ft. in diameter. The average length of each shoot was 20 ft. and many of these gave rise to secondary branches. If only the number of primary shoots were considered, with their average length of 20 ft., this plant would have a vine with a linear dimension of 7,200 ft., all developed in a five month growing season. Counts were made of the number of leaves on 100 shoots, and the total number for this plant was calculated to be 15,720 (Dittmer and Talley, 1964, p. 122).

A single buffalo gourd root may weigh up to 88 pounds after three or four seasons of growth and may contain nearly 20 percent starch, which can be extracted for use as a sweetener, or stabilizer. This starch also could be used to make puddings similar to tapioca, which is made from manioc (Hinman, 1984, p. 1447). The buffalo gourd is a drought-resistant plant whose seeds contain large amounts of oil and protein. Average values for crude oil and crude protein from buffalo gourd seed were 30 and 32 percent, respectively (Bemis et al., 1978, p. 91).

CULTIVATION

Buffalo gourd is on the threshold of being developed as a potential food crop for arid and semiarid regions in the United States and the Third World (Hinman, 1984, p.

1447). For planting on a field scale, results of research at the University of Arizona suggest direct seeding of recently developed hybrid seeds with initial "in-row" spacing of 6 to 12 inches, and rows 40 to 80 inches apart. At the end of the first season of growth, the fruits can be harvested mechanically. After the second-season harvest, alternate rows can be dug for their roots. These blank rows would then be planted with pieces of rooted vines from the adjacent rows. This procedure would continue as long as the vines remained productive. Yields are estimated at 12,000 lbs/acre of starch, 940 lbs/acre of oil, and 400 lbs/acre of protein (ibid., p. 77).

Buffalo gourd can be propagated easily by seed or root cuttings. Only a small piece of the root crown is needed to get it started. For use in a wild garden, its placement should be carefully considered because it will run over anything near it.

Dalea candida
White Prairie Clover

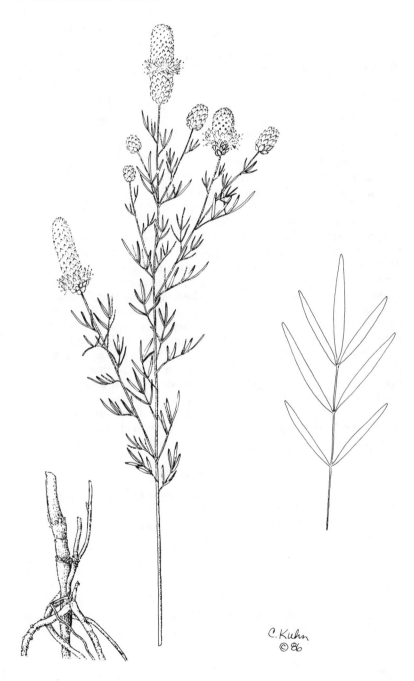

C. Kuhn
© 86

Prairie clover, white prairie clover, slender white prairie clover.

INDIAN NAMES

The Omaha and Ponca call the prairie clovers "makan skithe" (sweet medicine), a name they used for several plants; the Pawnee call them "kiha piliwus hawastat" (broom weed), because they used the tough stems as a broom to sweep their lodges, and also "kahts-pidipatski" (small medicine) (Gilmore, 1977, p. 42). The Lakota call purple prairie clover (*Dalea purpurea* Vent.) "toka'la tapejut'ta hu win'yula" (kit fox's medicine stem female), and the white prairie clover is its male equivalent, partly because it has coarser leaves (Rogers, 1980a, p. 72). The Kiowa name for white prairie clover is "khaw-tan-ee" (Vestal and Schultes, 1939, p. 33).

SCIENTIFIC NAME

Dal'ea can'dida Michx. ex Willd. is a member of the Fabaceae (Bean Family). *Dalea* is named in honor of Samuel Dale, an English botanist (1659–1739). The species name refers to flower color: *candida* means "of dazzling white."

DESCRIPTION

Perennial herbs from thick taproots, stems 1–several, 3–10 dm (12–40 in) tall, ribbed, sometimes dotted with glands. Leaves alter-

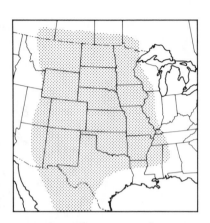

nate, pinnately compound, 1.5–6 cm (⅝–2⅜ in) long; leaflets in 3–5 pairs, egg-shaped to elliptic or oblong, surfaces with minute dots. Flowers in oval to cylindrical groups at tops of stems, from May to Sep; petals 5, white, upper 1 larger and erect, lower 2 boat-shaped, 2 wings at sides. Fruits dry, oval, 2.6–4.5 mm (1/16–3/16 in) long, dotted with glands, sepals persisting around bases.

Dalea purpurea differs in lacking glands and in having smaller leaves and purple flowers.

HABITAT

Prairies and rocky open woods.

PARTS USED

Roots—eaten raw; leaves (summer)—dried for tea.

FOOD USE

The prairie clovers are widely distributed and common, but only minor food sources. Their roots were chewed for their pleasant, sweet taste by many of the Indian tribes that lived on the prairies. The Comanche chewed on the roots of purple prairie clover (Carlson and Jones, 1939, p. 523); the Ponca ate roots of both white and purple species (Gilmore, 1977, p. 42); the Blackfoot ate white prairie clover roots (Johnston, 1970, p. 314); and the Lakota chewed on white prairie clover roots as we chew on gum (Munson, 1981, p. 237). The Kiowa also ate the roots of the white prairie clover, first peeling off the outer portion. This was one of their "ancient foods," but had been abandoned as a food source by the 1930s, which was probably due "as much to the difficulty of procuring and preparing them as to their coarse unpalatable nature" (Vestal and Schultes, 1939, p. 72).

The leaves of prairie clover were also dried and used for tea. The Blackfoot used the white prairie clover leaves (Johnston, 1970, p. 314) and the Dakota (Sioux) used the leaves of both white and purple prairie clovers (Gilmore, 1977, p. 42). The tea made from the leaves of purple prairie clover has been reported to produce a constipating effect (which is helpful for treating diarrhea) (Steyermark, 1981, p. 901). In *The Prairie Garden*, it is reported that the deep taproot also makes a fine-tasting tea (Smith and Smith, 1980, p. 135).

John Ernest Weaver, a prairie ecologist, reported (1934, p. 196) that the white and purple prairie clovers were the third most important legume and the eighth most important forb of upland prairies (Weaver, 1934, p. 196). The prairie clovers begin their growth in the spring with the prairie grasses but grow quickly, so that their conspicuous white or purple flowers are evident above the grasses in June or July.

CULTIVATION

The prairie clovers can be cultivated as tea plants and attractive ornamentals. They are easy to propagate by seed. Stratified seed, planted in the spring, will germinate and grow quickly the first year. For larger areas or prairie restorations, seeds can be sown in the fall. A pound of white prairie clover seed contains 384,000 seeds, and only about nine seeds per square foot are recommended for larger plantings (Salac et al., 1978, p. 4). White prairie clover grows rapidly after germinating, reaching 6–9 inches by midsummer. Blossoms may occur the first year and, unless there is an unusual drought, will appear the second. By midsummer the taproot will be 14 to 22 inches deep; the strong taproot of the mature white prairie clover extends downward 3.5–6 feet (Weaver and Fitzpatrick, 1934, p. 223).

Erythronium mesochoreum
White Dog's Tooth Violet

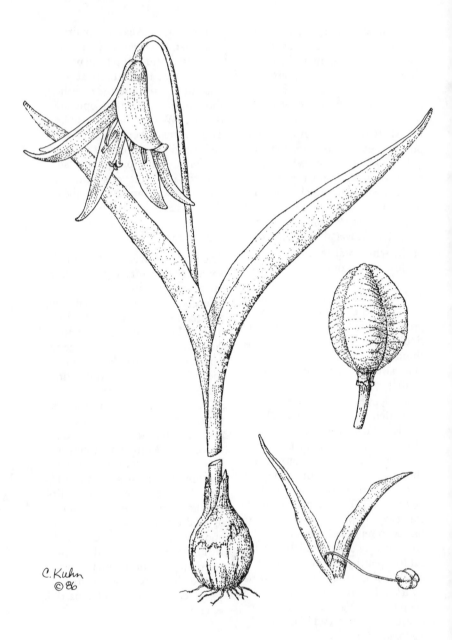

C. Kuhn
© 86

COMMON NAMES

White dog's tooth violet, prairie fawnlily, midland fawnlily, dog-tooth violet, spring lily, and snake lily.

INDIAN NAMES

None were found in the sources consulted.

SCIENTIFIC NAME

Erythron'ium mesochor'eum Knerr is a member of the Liliaceae (Lily Family). Linneaus derived *Erythronium* from the Greek word for "red", in reference to a red-flowered European species. The species name *mesochoreum* means "in the middle place," referring to its midcontinental (Iowa, Nebraska, Missouri, Kansas, and Oklahoma) distribution.

DESCRIPTION

Perennial herbs, growing from bulbs, no stems. Leaves 1 or 2, basal, lance-shaped, 8–15 cm (3⅛—6 in) long, shorter than flowering stalks, sometimes folded. Flowers solitary and nodding at tops of erect stalks in Mar and Apr; each 2.5–5 cm (1–2 in) long, with 6 segments, elliptic, spreading, white tinged with lavender and with yellow spots at bases. Fruits dry, egg-shaped, 1–4 cm (⅜–1⅝ in) long, resting on ground, opening into 3 sections.

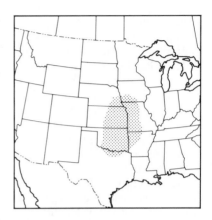

HABITAT

Prairie.

PARTS USED

Bulbs (late spring)—raw or cooked; leaves (spring)—cooked.

FOOD USE

The white dog's tooth violet is one of the earliest wildflowers of the spring. It is closely related to the dog-tooth violet or white fawn-lily, *E. albidum* Nutt., of the eastern woodlands. Like other dog's tooth violets, our prairie species has showy, lily-like flowers. The white dog's tooth violet is not a major food source. Moreover, it is not very common, so it is best not to harvest it from the wild.

The only reference to Indian use of this plant is by the Winnebago. It was reported that Winnebago children (apparently on their reservation in Iowa) ate the roots raw "with avidity," when dug in the springtime (Gilmore, 1977, p. 19).

Jacob Bigelow, an early American physician, claimed that the bulbs of the related fawn lily, *E. americanum* Ker., are emetic (Bigelow, 1794, as cited in Fernald et al., 1958, p. 132). However, more recent tests have shown no such property (ibid.). To my sense of taste, the bulbs of these two *Erythronium* species are almost identical, crunchy and mild, with a slightly bitter aftertaste.

Josiah Gregg was one of the leaders in opening the Santa Fe Trail and establishing trade with the Southwest. He made a rare observation of the white dog's tooth violet (most probably in eastern Kansas):

The flowers are among the most interesting products of the frontier prairies. The gay meadows wear their most fanciful piebald robes from earliest spring till di-vested of them by the hoary frosts of autumn. When again winter has fled, but before the grassy green appears, or other vegetation has ventured to peep above the earth, they are bespeckled in many places with a species of erythronium, *a pretty little lilaceous flower, which springs from the ground already developed, between a pair of lanceolate leaves, and is soon after in full bloom (Moorhead, 1954, p. 363).*

CULTIVATION

Julian Steyermark (1981, p. 434) reported that the two closely related species, *E. albidum* and *E. americanum* can be successfully transplanted to a garden, but it may be a year or two before their showy flowers are seen. Once established, they will thrive in a wildflower garden and the same would be expected to be true for the white dog's tooth violet.

Fragaria virginiana
Wild Strawberry

C. Kuhn
© 86

COMMON NAMES

Wild strawberry and strawberry.

INDIAN NAMES

The Pawnee name is "aparu-huradu" (berry, ground); the Winnebago name is "haz-shchek," and "haz" means fruit; the Omaha and Ponca called it "bashte"; and the Dakota called it "wazhush-teca"; no translations provided (Gilmore, 1977, p. 32). The Lakota (Sioux) name, which is very similar and perhaps the same as the above Dakota (Sioux) name, is "wazi'skeca" (pine mink) (Rogers, 1980b, p. 56). The Cheyenne called it "ve'shkee'?ehe-menoste" (sweet berries) (Hart, 1981, p. 34). The Blackfoot name is "oht-tchis-tchis" (Johnston, 1970, p. 313) and the Osage name is "ba-stse'-ga" (La Flesche, 1932, p. 337); neither name is translated.

SCIENTIFIC NAME

Fragar'ia virginia'na Duchesne is a member of the Rosaceae (Rose Family). *Fragaria* is from the Latin "fraga," the classical name used for the strawberry fruit, referring to its fragrance. The species name *virginiana* means "of Virginia."

DESCRIPTION

Perennial herbs with thick rhizomes, spreading and forming colonies by horizontal stems that root and produce plantlets. Leaves in rosettes, each with 3 leaflets, broadly elliptic or 4-angled, to 5

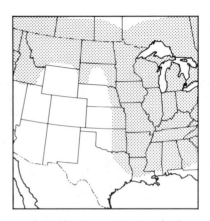

cm (2 in) long, margins toothed. Flowers in clusters on stalks shorter than leaves, from Mar to Jun; petals 5, separate, egg-shaped, 6–14 mm (¼–½ in) long, white. Fruits dry, small, appearing like seeds on the surfaces of fleshy, top-shaped, red, and juicy structures (popularly referred to as fruit), ripening in May to Jul.

HABITAT

Prairies, banks, roadsides, and openings in woodlands.

PARTS USED

Ripe fruit (late spring, early summer)—raw, cooked, or dried; young leaves (spring)—steeped for tea.

FOOD USE

Wild strawberries have a sweet, delicious flavor. All the Indian tribes of the Missouri River region "luxuriated in them in their season" (Gilmore, 1977, p. 32). The

Dakota even called their lunar month corresponding to June "Wazhushtecha-hu" (the moon when strawberries are ripe) (ibid.).

The ripe fruits were gathered and eaten fresh by the Hidatsa (Nickel, 1974, p. 64), Ponca (Howard, 1965, p. 44), Cheyenne (Hart, 1981, p. 35), and Omaha (Dorsey, 1881, p. 307). Melvin Gilmore (1977, p. 32) reported that wild strawberries are too moist to dry easily. However, the artist George Catlin, after visiting the Mandan in 1832 (1973, p. 122) stated that: "Great quantities of wild fruit of different kinds are dried and laid away in Store for the winter season, such as buffalo berries, service berries, strawberries, and wild plums." Also, the Pawnee were reported to use them fresh and dried for flavoring other dishes (Dunbar, 1880, p. 324).

The Blackfoot and the Winnebago also used strawberry leaves for tea (Johnston, 1970, p. 313; Gilmore, 1977, p. 32). The strawberry was important to the Blackfoot and influenced their perception of the world, indicated by their name for the Hand Hills in Alberta, "Oht-tchis-tchis," which means "Strawberry Hills" (Johnston, 1970, p. 313).

Fruits on the prairie were greatly appreciated by early travelers and settlers. In 1844, Josiah Gregg, a promoter of the Santa Fe Trail, published the following in his *Geography of the Prairies:* "With regard to fruits, the prairies are of course not very plentifully supplied. West of the border, however, for nearly two hundred miles, they are covered, in many places with the wild strawberry" (Moorhead, 1954, p. 363).

Thomas Farnham, after traveling about the first 25 miles over the Santa Fe Trail, reported in 1839 from Elm Grove or Round Grove (in present-day northeast Kansas): "At this encampment final arrangements were made for our journey over the Prairies. . . . Officers were also chosen and their powers defined; and whatever leisure we found from these duties during a stay of two days, was spent in regaling ourselves with strawberries and gooseberries, which grew in great abundance near our camp" (Thwaites, 1906, 28: 54).

Wild strawberries are smaller and more flavorful than cultivated varieties. As one naturalist wrote: "I had rather have one pint of wild strawberries than a gallon of tame ones" (Medsger, 1966, p. 21). They are also nutritious, having more vitamin C than an equal weight of oranges (Watt and Merrill, 1963, p. 60).

CULTIVATION

Ripe, cultivated strawberries from the garden with their delightful flavor are a special addition to any meal. The history of strawberry cultivation starts in America. Roger Williams, an early colonist of Rhode Island, said in 1643: "This berry is the wonder of all the fruits growing naturally in these parts. It is of itself excellent; so that one of the chiefest doctors in England was wont to say, that God could have made, but God never did make, a better berry. In some parts where the Indians have planted, I have many times seen as many as would fill a good ship, within few miles compass" (Hedrick, 1919, p. 282).

The common, cultivated strawberry that is grown commercially and in gardens is considered to be of hybrid origin between our wild strawberry, *F. virginiana*, and the Chilean *F. chiloensis* (L.) Duchesne. It was first developed in France and has been much improved by American breeders so that it is grown over a wider distribution than any other temperate-zone fruit. Wild strawberries can be transplanted in early spring and are identical to the cultivated varieties in their cultural requirements.

Glycyrrhiza lepidota
American Licorice

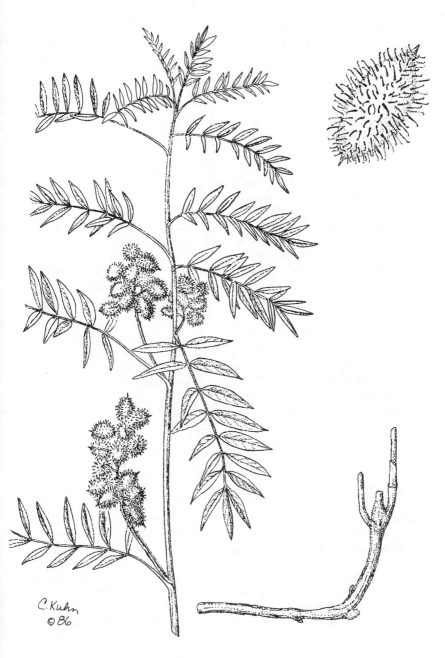

C. Kuhn
© 86

American licorice, wild licorice, licorice, and dessert root.

The Pawnee name is "pithahatu-sakitstsuhast" (Gilmore, 1977, p. 40) or "pilahatus" (Gilmore, 1914, p. 4), not translated. The Dakota name is "wi-nawizi" (jealous woman) in reference to the burrs, which "take hold of a man" (Gilmore, 1977, p. 40). The Cheyenne name is "haht' nowassoph" (yellow-jacket stinger plant) referring to the color and the burrs, which stick to a person like a yellow-jacket or wasp (Grinnell, 1962, p. 178).

Glycyrrhi'za lepido'ta Pursh is a member of the Fabaceae (Bean Family). *Glycyrrhiza* comes from Greek and means "sweet root." The species name *lepidota* means "scaly" since the leaves have minute scales when young.

Perennial herbs from deep, woody rhizomes; stems 3–10 dm (12–40 in) tall, branching, dotted with glands. Leaves alternate, pinnately compound; leaflets 7–21, oblong to lance-shaped with sharp points at tips, up to 5 cm (2 in) long, surfaces scaly when young and dotted with glands later. Flowers in narrow, elongated groups among leaves, from May to Aug; petals 5,

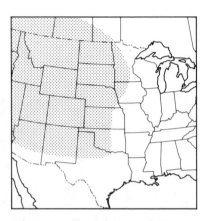

white or yellowish, top 1 larger and erect, lower 2 boat-shaped, 2 wings at sides. Fruits dry, oblong, 1–2 cm (⅜–¾ in) long, covered with hooked prickles; seeds olive green to brown, smooth.

Prairies, pastures, waste places, and along railroad rights-of-way.

Roots (fall to winter)—edible raw or cooked, and also dried for a tea.

American licorice is native to the Prairie Bioregion and has a characteristic sweet, licorice-tasting substance in its roots, which is used for food and medicinal purposes. Captain Meriwether Lewis of the Lewis and Clark expedition reported, on July 20, 1806, seeing an abundance of wild licorice on the plains and river bottoms near the Milk River (in present-day Montana). Later during the expedition, the Indian method of preparation was recorded. The roots were roasted in embers, then pounded with a stick in order to separate the tough, woody string from the center of the root. This string was removed, leaving a food that had a taste similar to that of sweet potatoes (Thwaites, 1905, 5: 213; Hart, 1976, p. 35).

The Cheyenne ate the shoots of licorice early in the spring when they were young and tender. They were eaten raw and reportedly were good until they were about a foot high and began to leaf out (Grinnell, 1962, p. 178).

In August of 1914, Melvin Gilmore, while Curator of the Nebraska State Historical Society Museum, accompanied Chief White Eagle of the Pawnee Indian Reservation in Oklahoma to one of the Pawnee's previous village sites along the Loup River in Nebraska, near present-day Genoa. Chief White Eagle, with the help of an interpreter, informed Gilmore (1914, p. 4) that "the Pawnee call this location 'Kitspilahatus' from its position on the creek which they call 'Kits'Pilahatus' from 'pilahatus,' the Pawnee name of *Glycyrrhiza lepidota* which grows there." The Pawnee names show the awareness that they had for the licorice plant and its importance to them.

During the summer of 1819, the Steven Long Expedition had passed the Pawnee villages and was heading west along the Platte River, nearing where it forks. Edwin James, the botanist of the expedition, reported seeing American licorice: "Among other plants observed about our encampment, was the wild liquorice . . . which is believed to be the plant mentioned by Sir A. Mackenzie, which is used as food by the savages of the north-west. The root is large and long, spreading horizontally to a great distance. In taste it bears a very slight resemblance to the liquorice of the shops, but is bitter and nauseous" (Thwaites, 1905, 15: 231).

Dr. Edward Palmer wrote in 1878 (p. 653) concerning American licorice that the "Pah-Utes eat it for its tonic effects." He also noted that it is called the dessert root by the settlers because it tastes much like licorice and is sometimes chewed in the place of tobacco.

Julian Steyermark, in the *Flora of Missouri* (1981, p. 911), reported that "the sweet, licorice-flavored root (of *Glycyrrhiza lepidota*) was eaten raw or in baked form by Indians. It is sometimes chewed by country folk for its pleasant taste. The species is related to the licorice of commerce (*G. glabra* L.), from the root of which is obtained an extract used in medicine as a laxative and expectorant, and in brewing, confectioneries, and flavoring tobacco."

In an analysis of the American licorice, glycyrrhizin (the active substance that gives the licorice taste) content was found to be higher in Nebraska roots than in two samples of roots from the state of Washington, but none of the three samples "possessed the sweet taste characteristic of the official plant (*Glycyrrhiza glabra* L.) traditionally grown in southern Europe and southwest Asia (Yanovsky and Kingsbury, 1938, p. 658).

There seems to be quite a variation in the flavor of American licorice. The roots I have sampled in central Kansas and eastern Colorado were bitter and medicinal tasting, with only a slight tinge of the sweet licorice flavor. Their flavor seemed to increase the longer they were chewed. It is possible that a different soil, a different location, or harvest at a different time could yield better tasting roots. However, I harvested the roots in the early winter, when they should have a high concentration of starches and sugars.

CULTIVATION

Oliver Perry Medsger, in *Edible Wild Plants* (1966, p. 199), reported that he found American licorice growing near Indian villages in New Mexico as though it had once been cultivated.

American licorice can be planted in a wild food garden, but it is known as a weedy and somewhat aggressive, spreading plant. Michael Moore, in *Medicinal Plants of the Mountain West* (1979, p. 97), reports that it can be propagated by transplanting, in early spring or late fall, pieces of the underground runners (approximately 8 to 10 inches long) that have scalelike buds, to a site that is moist and well drained. They should be planted about 4 inches deep. "With the right kind of growing conditions and a five-year headstart, this stout, fast-spreading plant will probably uproot the nearest house" (ibid.).

Helianthus annuus
Sunflower

C. Kuhn
©86

COMMON NAMES

Sunflower, common sunflower, Kansas sunflower, and mirasol.

INDIAN NAMES

The Kiowa name is "ho-son-a" (looking at you) (Vestal and Schultes, 1939, p. 60). The Dakota name is "wahcha-zizi" (yellow flowers); the Omaha and Ponca name is "zha-zi" (yellow weed), and the Pawnee name is "Kirik-tara-kata" (yellow eyes) (Gilmore, 1977, p. 78).

SCIENTIFIC NAME

Helian'thus ann'uus L. is a member of the Asteraceae (Sunflower Family). *Helianthus* comes from the Greek *helios anthos*, meaning "sun flower." The species name *annuus* means "annual."

DESCRIPTION

Annual herbs 6–25 dm (2–10 ft) tall, stems hairy, branching at top. Leaves mostly alternate, heart-shaped to spade-shaped, 10–40 cm (4–16 in) long, surfaces rough, margins usually toothed. Flower heads at ends of branches, from Jul to Sep; ray florets 17 or more, at least 2.5 cm (1 in) long, yellow, disk florets numerous, reddish to purple. Fruits dry, egg-shaped, 3–5 mm (⅛–³⁄₁₆ in) long, smooth with 2 small bristles at top, ripening in Sep to Oct.

HABITAT

Waste and cultivated ground, low meadows, prairies, along roadsides and railroads.

PARTS USED

Fruits, popularly called seeds (fall)—raw, cooked, roasted, or dried and ground, and as a source of oil; flower buds (summer)—boiled; roasted shells or seeds for a coffee substitute.

FOOD USE

The sunflower was domesticated by the Indians, and is an important agricultural crop and a widespread weed. It has an extensive history as a wild food plant. When Lewis and Clark were along the Missouri River in Montana, the following entry was made in their journal for July 17, 1805:

Along the bottoms, which have a covering of high grass, we observe the sunflower blooming in great abundance. The Indians of the Missouri, more especially those who do not cultivate maize, make great use of the seed of this plant for bread, or in thickening their soup. They first parch and then pound it between two stones, until it is reduced to a fine meal. Sometimes they add a portion of water, and drink it diluted; at other times they add a sufficient proportion of marrow-grease to reduce it to the consistency of common dough, and eat it in this manner. This last composition we preferred to all the rest, and thought it at that time a very palatable dish (Thwaites, 1904, 2: 238).

Edward Palmer stated in "Report of the U.S. Commission of Agriculture" (1871, p. 419) that the seeds of sunflowers are often gathered in the West where they are found growing "on river bottoms and rich moist spots on the prairies, . . . Being very sweet and oily, they are eaten raw, or pounded up with other substances, made into flat cakes and dried in the sun, in which form they appear to be very palatable to the Indians."

The Mescalero and Chiricahua Apache made extensive use of wild sunflowers. They harvested both annual species, *H. annuus* and *H. petiolaris* Nutt., in the fall. Apache women harvested the ripe seeds of the former species in the early 1930s in the following manner. They "placed a basket under the plant and sharply struck the back of the sunflower head with a stick, knocking the seeds into the basket. They were sometimes ground and the flour used for a thick gravy, but more commonly this was sifted, made into dough, and baked on hot stones or in hot ashes. This kind of bread is still in common use" (Opler, 1936, p. 48).

The Hidatsa cultivated sunflowers, but they also used wild ones, which were preferred because the seeds of their smaller flower heads produced a superior oil. The heads were gathered late in the season after being frosted, which made them oilier (Wilson, 1917, pp. 18–19).

The seeds of wild sunflowers are smaller and hard to shell, but taste as good as the domesticated varieties. They can be eaten raw, boiled, or roasted. The whole seeds or just their shells can be roasted, ground, and used as a coffee substitute. Edward Palmer (1878, p. 602) described the preparation of one tasty Indian food made of flour of wild sunflower kernels. "The meal or flour is also made into thin cakes and baked in hot ashes. These cakes are of a gray color, rather coarse looking, but palatable and very nutritious. Having eaten of the bread made from sunflowers I must say that it is as good as much of the corn bread eaten by Whites."

Wild sunflowers were readily available as food for Indians in the prairies. Their abundance in the region was reported by the botanist John Torrey, who received botanical specimens and a detailed report from Lieutenant Frémont following his 1842 expedition to the Rocky Mountains: "The valley of the North fork (of the Platte River) is without timber; but the grasses are fine, and the herbaceous plants abundant. On the return of the expedition in September, Lieut. Frémont says the whole country resembled a vast garden; but the prevailing plants were two or three species of *Helianthus* (sunflower)" (Jackson and Spence, 1970, p. 287).

The importance of sunflowers to Native Americans is further indicated by the widespread presence of sunflowers in their mythic belief system. In the Northeast, they are part of the Onandaga (Iroquois) creation myth (Gilmore, 1977, p. 78). In the Southwest, the Hopi believed that when the sunflowers are numerous, it is a sign that there will be an abundant harvest (Whiting, 1939, p. 97). In the prairies, the Teton Dakota had a saying: "when the sunflowers were tall and in full bloom, the buffalos were fat and the meat good" (Gilmore, 1977, p. 78).

Sunflowers were introduced into England, and John Gerarde, a herbalist, reported in the 1630s an unusual food use of sunflowers: "We have found by triall, that the buds before they be flowered, boiled and eaten with butter, vine-gar and pepper, after the manner of artichokes, an exceeding pleasant meat, surpassing the artichoke far in procuring bodily lust" (Gerarde, 1633, in Hedrick, 1919, p. 298).

The abundance of sunflowers in the prairies, along with their beauty and symbolism, prompted the legislature of Kansas in 1903 to proclaim the wild native sunflower as the state flower and floral emblem (Bare, 1981, p. 445).

CULTIVATION

The sunflower is a native domesticated crop. During the last 3,000 years, Indians increased the seed size approximately 1,000 percent. They gradually changed the genetic composition of the plant by repeatedly selecting the largest seeds (Yarnell, 1978, p. 297).

The boundary between wild and domesticated sunflowers is not clear. It is believed that the wild sunflower originated in the general region of the Colorado Plateau. However, "modern sunflowers are abundantly represented throughout the greater Midwest and in the Great Plains by large ruderal weed forms growing in highly disturbed habitats. It is possible that they represent an ancestral type intermediate between wild and cultigen varieties, but it is just as likely that the ruderal forms are feral descendents of early domesticated forms" (ibid., p. 291). Charles Heiser, in *The*

Sunflower among North American Indians (1951, p. 436) states that the original source of the cultivated sunflower is not known, but one possibility is from Indian agriculturalists along the Missouri River in the Prairie Bioregion.

Archaeological evidence supports this hypothesis and also shows the antiquity of the use of sunflowers as food and the gradual increase in size of their seeds. A sunflower specimen, no larger than a wild one, was found at the Koster site along the Illinois River, in southwest Illinois, and has been dated to 5800 B.C. (Asch and Asch, 1982, p. 15). Considerable quantities of sunflower seeds were found in human feces that were well preserved in the dry caves of Newt Kash Hollow and Salts Cave in Kentucky (Yarnell, 1978, p. 292). They were larger than seeds of modern wild sunflowers, which indicates that the plants were actually being cultivated. Specimens from Salts Cave were dated from 1500 to 300 B.C. (ibid.). Seeds from the Mitchell site, in southeast South Dakota (Benn, 1974, p. 225), and the Two-Deer site in Butler County of eastern Kansas (Adair, 1984, p. 121) show that sunflowers were being cultivated in these locations around 1000 A.D.

The Mandan, Arikara, and Hidatsa were known to be cultivating sunflowers in historic times. Prince Maximilian reported in 1832 that the Mandan were cultivating "two or three varieties of sunflowers with red, and black, and one with smaller seeds. Very nice cakes are made of these seeds" (Thwaites, 1906, 23: 275). Gilbert Wilson in 1917 (pp. 17–18) gave one of the most complete accounts of sunflower cultivation. He interviewed Buffalobird Woman, an elderly Hidatsa, who was an expert agriculturalist and reported to him:

The first seed that we planted in the spring was the sunflower seed. Ice breaks on the Missouri about the first week in April; and we planted sunflower seed as soon after as the soil could be worked. Our native name for the lunar month that corresponds most nearly to April, is "Mapi'-o' ce-mi'di" or Sunflower-planting-moon.

Planting was done by hoe, or the woman scooped up the soil with her hands. Three seeds were planted in a hill, at the depth of the second joint of a woman's finger. The three seeds were planted together, pressed into the loose soil by a single motion, with thumb and first two fingers. The hill was heaped up and patted firm with the palm in the same way as we did for corn.

Usually we planted sunflowers only around the edges of a field. The hills were placed eight or nine paces apart; for we never sowed sunflowers thickly. We thought a field surrounded thus

by a sparce-sown row of sunflow-
ers, had a handsome appearance.
. . . Of cultivated sunflowers we
had several varieties, black,
white, red, striped. . . . The varie-
ties differed only in color.

Each family raised the variety
they preferred. The varieties were
well fixed; black seed produced
black; white seed, white.

Our sunflowers were ready for
harvesting when the little petals
that covered the seeds fell off, ex-
posing the ripe seeds beneath.
Also, the back of the head turned
yellow; earlier in the season it
would be green. . . . Some of these
big heads had each a seed area as
much as eleven inches across. Be-
sides these larger heads, there
were other and smaller heads on
the stalk; and wild sunflowers
bearing similar small heads grew
in many places along the Mis-
souri, and were sure to be found
springing up in abandoned gar-
dens.

To harvest the larger heads, I
put a basket on my back, and
knife in hand, passed from plant
to plant, cutting off each large
head close to the stem; the sev-
ered heads I tossed into my bas-
ket. These heads I did not let dry
on the stalk, as birds would de-
vour the seeds.

My basket filled, I returned to
the lodge, climbed the ladder to
the roof, and spread the sunflower
heads upon the flat part of the
roof around the smoke hole, to
dry. . . .

When the heads had dried
about four days, the seeds were
threshed out; . . .

To thresh the heads, a skin was
spread and the heads laid on it
face downward, and beaten with
a stick.

Today, the sunflower is an important alternative crop in the
Midwest for production of oil and
edible seeds. Commercially available varieties with very showy
flower heads can be grown in
sunny areas of a garden. The seeds
will attract birds in autumn.

Helianthus tuberosus
Jerusalem Artichoke

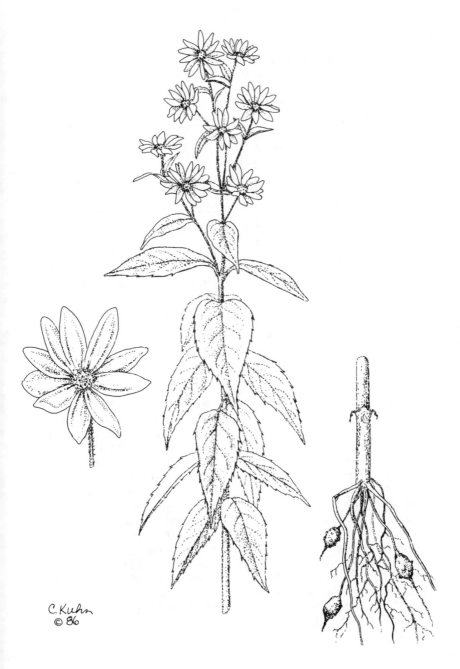

C. Kuhn
© 86

COMMON NAMES

Jerusalem artichoke, tuberous sunflower, sunchoke, and girasole. The common name Jerusalem artichoke is misleading because the plant is neither from Jerusalem nor is it an artichoke (Fernald et al., 1958, p. 359). A translation of the scientific name, "tuberous sunflower," would be more appropriate.

INDIAN NAMES

The Cheyenne name is "hohinon" (brought back by scouts) (Grinnell, 1962, p. 189). The Pawnee name is "kisu-sit" (tapering, long) (Gilmore, 1977, p. 79).

SCIENTIFIC NAME

Helian'thus tubero'sus L. is a member of the Asteraceae (Sunflower Family). The scientific name means "sunflower with tuberous swellings of its rhizomes."

DESCRIPTION

Perennial herbs from tuber-bearing rhizomes, 1–3 m (3–10 ft) tall, branching near tops. Leaves mostly opposite, but upper ones alternate, egg-shaped to lance-shaped, 10–25 cm (4–10 in) long, surfaces hairy, margins usually coarsely toothed. Flower heads at ends of branches from Aug to Oct; ray florets 10–20, up to 4 cm (1⅝ in) long, yellow, disk florets yellow. Fruits dry, egg-shaped, 5–7 mm (³⁄₁₆–¼ in) long, hairy.

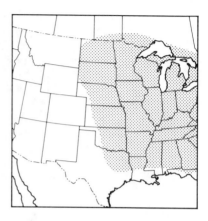

HABITAT

Wet places in prairies, open woods and thickets, stream margins, disturbed habitats, and roadsides.

PARTS USED

The roots (fall, winter, or early spring)—raw or cooked.

FOOD USE

The Jerusalem artichoke is a productive plant that is native to the eastern half of North America. Its tubers were widely used by the Plains Indians and also by many eastern woodland tribes. Gilmore (1977, p. 79) reported that the Pawnee only ate the roots raw, whereas the Omaha, Ponca, Winnebago, and Dakota ate them raw, boiled, or roasted. While it supplemented the cultivated crops (corn, beans, and squash) in their diet, none of these tribes claimed to have ever cultivated the plant (ibid.).

The Omaha ate the tubers of this sunflower in the early spring. This was the time when the winter diet of stored foods had become monotonous or, in a difficult year, when food was in short supply. In either case, an available fresh food was appreciated. The Omaha spoke of the roots as "the food of homeless boys who had no relative to feed them" (Fletcher and La Flesche, 1911, p. 341).

Captain Meriwether Lewis of the Lewis and Clark expedition, when along the Missouri River, in present-day North Dakota, reported on April 9, 1805:

When we halted for dinner the squaw (Sacajawea) busied herself in serching for the wild artichokes which the mice collect and deposit in large hoards. This operation she performed by penetrating the earth with a sharp stick about some small collections of drift wood. Her labour soon proved to be successful, and she procured a good quantity of these roots. . . . The root is white and of ovate form, from one to three inches in length and usually about the size of a man's finger. One stalk produces from two to four, and somitimes six of these roots (Thwaites, 1904, 1: 290).

Domesticated Jerusalem artichokes sometimes produce yields exceeding those of potatoes. The roots of wild tuberous sunflowers are not always easy to find, and in a competitive or dry environment, few tubers are produced. Like many other wild foods, the wild tubers are smaller but more flavorful than their domesticated relatives. Tubers are not produced until the late summer and fall, when the length of days has shortened considerably.

Tubers should not be dug until after frost, and waiting until spring is fine, even preferable, because they become sweeter with age. This occurs because the tubers are primarily composed of water and the carbohydrate inulin, which is changed to fructose so growth can begin in the spring (Stevens, 1961, p. 379). The uncooked tubers are crunchy and taste like water chestnuts. They are good in salads. When cooked, the tubers are sweeter than potatoes, although not so firm.

The tubers of this native sunflower are nutritious and healthful, being high in iron but low in fat and available carbohydrates. The caloric content of a 100-gram portion of the raw tubers varies from seven calories for freshly harvested tubers to 75 calories for those stored for a long period (Watt and Merrill, 1963, p. 34). Possible explanations of this increase in caloric content include the loss of moisture and the change of starch to sugar over time.

The nature of the carbohydrate content of Jerusalem artichokes is one reason why some people like them and others don't. Jerusalem artichokes are frequently recommended as a good food for weight watchers or diabetics because they contain inulin, a carbohydrate source that is largely indigestible. In other words, only small amounts of carbohydrates are available for digestion, even though a lot of bulk has been ingested. However, the indigestible nature of inulin affects the body in a way reported by the Dakota in the Minnesota Territory. They ate the tubers only "when in a state of starvation, from dread of its flatulent qualities" (Prescott, 1849, p. 452). Alexander Henry, an explorer, also reported: "When boiled, the root is tolerably good eating but, when eaten raw, is of a windy nature and sometimes causes a severe colic" (Kaldy et al., 1980, p. 352). The slow baking procedure described for *Camassia scilloides* probably would render sunflower tubers sweeter and more digestible.

CULTIVATION

The Jerusalem artichoke is native to the prairies; Asa Gray, an eminent early American botanist, believed that it originated in the Mississippi Valley. In a letter to Gray concerning cultivation by Native Americans, J. Hammond Trumbull wrote on March 26, 1877: "The historical evidence that 'artiscohoki sub terra' were cultivated in Canada and in some parts of New England before the coming of Europeans is tolerably

clear. The only question, if there be any, is as to species, and this does not appear to have been raised for more than half a century after the 'Jerusalem artichoke' was known to English and Continental botanists" (Trumbull and Gray, 1877, pp. 348–349).

Richard Ford, an ethnobotanist at the University of Michigan, has stated that although the Jerusalem artichoke was encouraged through habitat manipulation by Native Americans, it was probably not a crop plant in the same sense that corn, beans, and squash were (Ford, 1980, personal communication, cited in Turner, 1981). But Charles Heiser, a botanist at Indiana University, has recently contended that the Jerusalem artichoke was being cultivated when Europeans arrived and that it may have been domesticated (1985, p. 2). In either case, it has not been totally domesticated and is not dependent upon humans for its continued existence. It is a very hardy and persistent food plant and has many of the characteristics (being a high-yielding perennial crop) that would make it suitable for use in a sustainable, low–maintenance agricultural system. It is not the ideal perennial crop, because its production comes from its roots, which would require an annual disturbance of the soil for harvest.

The Jerusalem artichoke is cultivated as a minor vegetable crop, and there have been some recent experiments with field-scale cultivation for livestock feed. It is easy to grow. Besides the tasty wild species, there are improved varieties that have been selected for their added productivity and whiteness. The tubers can be planted in the fall or spring—but one should first consider if one's grandchildren will want them there. Jerusalem artichokes can be controlled but it takes considerable work. Although any sunny location is suitable, I have found that a corner of the garden or even a place outside the garden is best for a future patch of tuberous sunflowers.

The tubers can be harvested in the late fall, winter, or spring before the soil becomes warm enough for the sprouts to emerge. The tubers often spread two or more feet away from the mother plant when growing in soft, rich soil. Usually, not all of them are found, so enough remain to start next year's crop.

Jerusalem artichokes are a welcome addition to the winter diet, but cannot be dug from frozen ground. When stored in the refrigerator for more than a week, they lose their crispness. I have stored my winter supply of tubers and roots as follows: In the fall, after a hard freeze, I dig up my tuberous sunflowers and bury them in a leaf-insulated pit about one and one-half to two feet deep, along with other root crops—such as turnips, Japanese burdock roots, parsnips, and carrots. I cover these tightly with more leaves, place a large board on top of the entire pit, and then place some plastic sacks full of leaves on top of the board.

Even in the dead of winter, with snow covering the frozen ground, I can go out to my cache pit for some crisp Jerusalem artichoke tubers to add to my supper. To keep mice and other small mammals from plundering your cache, as once happened to mine, pick a location where there has been little sign of their activity and that is free from vegetation suitable for mouse nests.

Ipomoea leptophylla
Bush Morning Glory

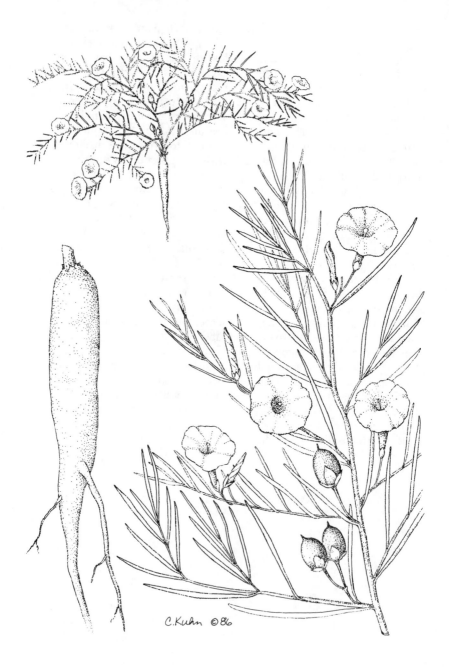

C.Kuhn ©86

COMMON NAMES

Bush morning glory, big-root morning glory, man root, man-of-the-earth (these names refer to the fact that the plant has man-sized roots), bush moonflower, and wild potato vine.

INDIAN NAMES

The Pawnee name for the bush morning glory is "kahts-tuwiriki" (whirlwind medicine). The Pawnee used the root for making a smoke-treatment medicine that alleviated nervousness and the effect of bad dreams. The plant reminded them of a whirlwind because of its twisted fibrovascular system (Gilmore, 1977, p. 58).

SCIENTIFIC NAME

Ipomoe'a leptophyl'la Torr. is a member of the Convolvulaceae (Morning Glory Family). *Ipomoea* means "worm-like" and *leptophylla* means "fine, narrow leaf."

DESCRIPTION

Perennial herbs from enlarged roots, stems lying on ground or erect, 0.3–1.2 m (12–48 in) tall. Leaves alternate, narrowly lance-shaped to linear, 3–15 cm (1¼–6 in) long, margins entire. Flowers solitary or in groups of 2–3 on long stalks among leaves, from May to Sep; petals fused into funnel-shaped tubes, flaring at tops, 5–9 cm (2–3½ in) long, purple-red to lavender-pink with darker throats. Fruits dry, oval, 1–

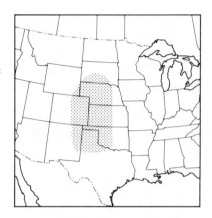

1.5 cm (⅜–⅝ in) long; seeds oblong, covered densely with short, brown hairs.

HABITAT

Sandy prairie, wasteground, roadsides, and banks.

PARTS USED

Root (fall to winter)—cooked.

FOOD USE

Bush morning glory is related to the sweet potato and has an extremely large, edible root. The Pawnee reportedly used it as a food source, "affording a partial subsistence at certain seasons" (Dunbar, 1880, p. 323).

Dr. Edward Palmer was an explorer, scientist, and adventurer of the American western frontier. In a report (1871, p. 407) he states the following about the bush morning glory: "This showy plant of the dry deserts of the West is commonly called man root, or man of the earth, being similar in size and shape to a man's body. The Cheyenne, Arapahoes, and Kiowas roast it for food when pressed by hunger, but it is by no means palatable or nutritious. Its enormous size and depth make its extraction by the ordinary Indian implements a work of much difficulty."

James Malin in *The Grassland of North America* (1961, pp. 446–7) described the bush morning glory and one of its first historical references:

It produced a root the weight of 10 to 100 pounds. Lieutenant Abert described his experience with it in 1846 while waiting for high water of the Pawnee Fork to subside, in the general vicinity of Larned, Kansas, about 99 degrees west longitude, in the hard land north of the Arkansas River. A soldier spent several hours trying to dig up a specimen under Abert's direction, but the ground was so hard they finally gave up and broke it off. The stem, about half an inch in diameter, ran down about 12 inches, then enlarged suddenly to 21 inches in circumference, or about 6 inches in diameter, and extended about 2 feet deeper. Abert's comment indicated that this specimen was relatively small compared with others supposed to grow to the size of a man.

John Ernest Weaver, who conducted field studies of prairie plant roots, reported (1956, p. 184) that some bush morning glory roots in sandy soil were 18 to 24 inches thick. The "lateral spread was enormous, the roots running off to distances of 15 to 25 feet or more in various directions from the base of the plant." The total depth of the roots could not be determined because of caving in of the sandy banks of the hole. However, "because of their diameters of 2 to 4 mm. at the 11-foot level and the nature of other roots examined, it is highly probable that they penetrated many feet deeper. The enlarged portion of the taproot furnishes not only an enormous reservoir of food but also a storehouse of water upon which the plant may exist during drought."

A closely related species, bigroot morning glory, *I. pandurata* (L.) G. F. W. May., also has a large edible root that was a food source of the Native Americans. It is a vine found on the eastern edge of the Prairie Bioregion, primarily in disturbed habitats. It is reported that the raw root is a purgative and so it is eaten only after being cooked. Bush morning glory also should be cooked before eating for the same reason.

Bush morning glory has been variously described as having an excellent taste to being nearly inedible (Elias and Dykeman, 1982, p. 216). It can be harvested at any time when the ground is not frozen, but is not as good in the summer when its starch reserves are low. Since the roots get woody and tough with age, they are best when harvested young. Roots can be boiled or baked. If the roots are bitter, boiling and changing the water two or three times are recommended. All roots have tough outer skins and require peeling (ibid.).

Fifty-seven carbonized morning glory seeds (*Ipomoea* species) were recovered from two excavated pits at Spoonbill, an early Caddo site in northeast Texas (Wood County) that was occupied between 800 and 1300 A.D. The presence of these seeds indicates that the residents were processing them for some purpose. Besides use for plant propagation, it is possible that the seeds were used to induce hallucinogenic visions, since they contain amides of lysergic acid that are related to LSD (lysergic acid diethylamide) (Crane, 1982, p. 86).

CULTIVATION

Because of its beautiful pink to purple flowers, bush morning glory has been suggested as an ornamental for roadsides, parks, and recreation areas. It is also recommended for planting in wildlife habitats and for prairie restorations within the central and western part of the Prairie Bioregion (Salac et al., 1978, p. 13). The bush morning glory could be developed as a cultivar for semi-arid climates because it is drought resistant, due to its extremely large, starchy root. It can be propagated from seed or by division of the root crown.

Iva annua
Marsh Elder

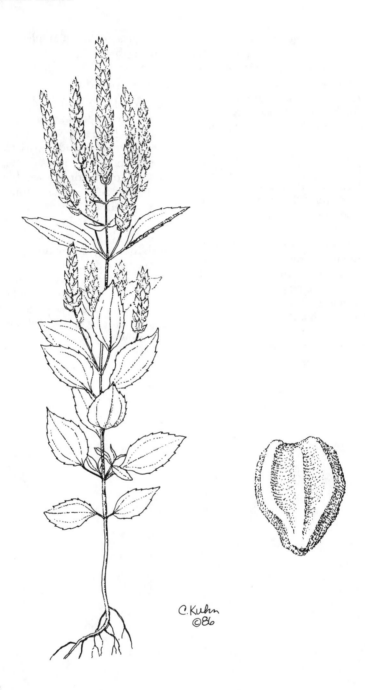

C. Kuhn
©86

Marsh elder, sumpweed, and samp.

The Lakota name for the closely related *Iva xanthifolia* Nutt. is "waxpe sica," which means "bad leaves" because, as one Lakota stated, "the seeds cause irritation to the bare skin" (Munson, 1981, p. 235).

I'va an'nua L. is a member of the Asteraceae (Sunflower Family). *Iva* is an ancient name for a medicinal plant. The species name *annua* means year, describing the lifetime of this plant.

Annual herbs, 4–20 dm (16–80 in) tall, upper stems hairy. Leaves mostly opposite, upper ones sometimes alternate, lance-shaped to egg-shaped, 5–15 cm (2–6 in) long, with 3 obvious veins, margins toothed. Flower heads in elongated groups at tops of branches, from Aug to Oct; florets small, greenish, no obvious rays. Fruits dry, 2–4 mm (¹⁄₁₆–³⁄₁₆ in) long, egg-shaped, and brown.

Waste ground; wet ground along streams, ponds, and sloughs; and river bottom prairies.

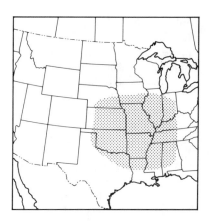

Fruits, popularly called seeds (fall)—probably roasted or cooked.

The use of marsh elder as food is not recommended because so little is known concerning how to process and eat it. It is included because its seeds (actually achenes) have been found extensively in archaeological remains in the east-central United States. In many cases, these archaeological specimens are up to four times larger than the seeds of the wild species today. This indicated to archaeologists that marsh elder was an ancient, domesticated crop.

The oldest record of the use of wild marsh elder seeds comes from the Koster site near St. Louis, which dates to about 5300 B.C. The seeds of the domesticated variety, *I. annua macrocarpa*, have been found in human coprolites in

Salts and Mammoths caves in Kentucky, which would appear to prove that this plant served as a food source (Asch and Asch, 1978, p. 331). At Mammoth Cave, which was occupied from 710 to 290 B.C., marsh elder seeds were identified in 87 of 100 preserved feces whose contents were examined.

The marsh elder is native from southern Illinois southward to Mississippi and northeastern Mexico and westward along river valleys into the prairies as far north as Nebraska (Yarnell, 1978, p. 291). However, the archaeological range of cultivated marsh elder varieties extends as far north as the Mitchell site in southeast South Dakota, which was occupied during the 11th century A.D. (Nickel, 1977, p. 56). The seeds were also found at the Yeo site near Kansas City, which was occupied from 635 to 870 A.D. and appears to have been a collection and storage station of hickory nut/marsh elder seeds for Hopewellian people (O'Brien, 1982, p. 37). An even earlier find in the Kansas City area is from the Trowbridge site, dated from 200 to 400 A.D. (Adair, 1984, pp. 33, 104).

Marsh elder seeds are similar in appearance to tiny sunflower seeds, but are unpalatable until they are processed because they have a tough outer shell and an objectionable odor and taste (Asch and Asch, 1978, p. 302). The Indians probably processed them during prehistoric times by roasting or boiling, which removes the objectionable flavor and odor, and also partially splits the outer shells. If they are then dried to reharden the kernels, rubbing will separate the kernels so the shells can be winnowed out (Asch and Asch, 1978, p. 302).

Wild marsh elder seeds are quite nutritious. They are oilseeds that contain almost 2,500 calories per pound and also contain several vitamins and minerals (Asch and Asch, 1978, p. 309).

Marsh elder has been reported to grow in some bottomland prairies, especially at the fluctuating margins of floodplain prairie lakes (Asch and Asch, 1978, p. 309). It is more commonly found in disturbed and often flooded habitats, sometimes in dense stands. Prehistoric Indians probably sought out these areas as harvest sites. Yields of wild marsh elder are reported by the anthropologists Nancy and David Asch (ibid., p. 313) to be "similar to other wild plants that were taken into cultivation as dietary staples."

CULTIVATION

The native people of the east-central United States developed the wild marsh elder into a new variety with larger seeds and greater yields. The size of the seeds increased about 1,000 percent (Yarnell, 1978, p. 297). It is believed that the cultivation of marsh elder probably began as an "element of necessity" to relieve a scarcity of storable food energy or to improve the reliability of the subsistence system so that even in years of drought or other adversity there would be enough food (Asch and Asch, 1978, p. 334). Marsh elder is believed to have been a stored food because in one specimen of the feces examined from Salts Cave, it occurred with strawberry fruits, which are seasonal (Yarnell, 1971, p. 554). Strawberries do not keep well and are difficult to dry. Marsh elder seeds ripen in the late fall, and to have been eaten with strawberries, they must have been stored for use in the spring.

The marsh elder was apparently abandoned as an agricultural crop before historic times, since it is not reported in any of the diaries and accounts written by early explorers or travelers. Why this happened will probably never be known, but it is believed that sunflowers, which were easier to cultivate, may have outdistanced the marsh elder in yield of seeds and oil (Yarnell, 1974, p. 340). Another possibility is that the development of a polycultural (mixed) cropping system in eastern North America, which began about 1000 A.D. with the introduction of the Mesoamerican squash, maize, and beans, was not compatible with the cultivation of marsh elder (Asch and Asch, 1978, p. 334). Marsh elder is similar to wheat in needing to grow in dense stands with high plant populations in order to produce a crop that is large enough to be harvestable. Like giant ragweed, marsh elder produces large amounts of pollen that is the bane of hay fever sufferers; this may also have encouraged its abandonment as a cultivated crop.

There is considerable concern within the scientific community about maintaining and expanding genetic diversity. If the cultivated variety of marsh elder had not been extinct by the time of European settlement, it is possible that it would have been improved even further and would be a field crop today. The marsh elder may have potential for further improvement and use as a crop plant in the future.

Liatris punctata
Gayfeather

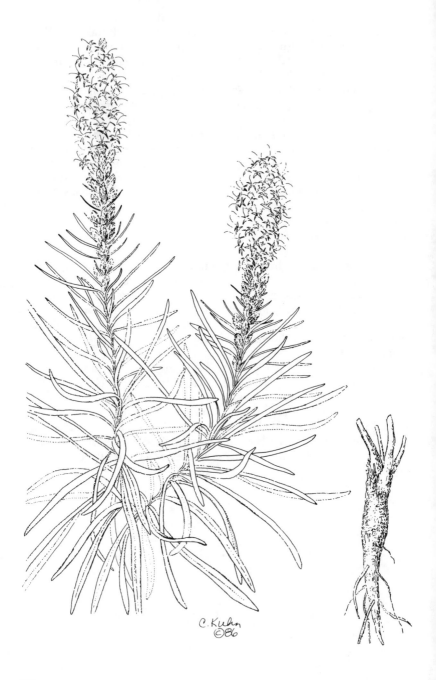

C. Kuhn
©86

Gayfeather, blazing star, dotted gay feather, Kansas gayfeather, dotted button-snake-root, and starwort. (These common names primarily refer to the showy flower stalk.)

INDIAN NAMES

The Omaha and Ponca names for gayfeather (*Liatris aspera*) are "aontashe" and "makan-sagi," which both mean "hard medicine" (Gilmore, 1977, p. 81). The Pawnee call this species "kahtsu-dawidu" (round medicine) (ibid.).

SCIENTIFIC NAME

Lia'tris puncta'ta Hook. is a member of the Asteraceae (Sunflower Family). The derivation of the name *Liatris* is unknown. The species name *punctata* is Latin for dotted, referring to the appearance of the leaves.

DESCRIPTION

Perennial herbs, 1–8 dm (4–32 in) tall, stems single or in clusters from rootstocks. Leaves alternate, linear, up to 15 cm (6 in) long, closely spaced, arching upwards, surfaces dotted. Flower heads as tufts arranged in cylindrical groups at tops of branches, from Jul to Oct; florets tubular, 9–12 mm (3/8–1/2 in) long, pink-purple, with 5 pointed lobes and long straplike styles protruding. Fruits dry, 10-ribbed, 6–7 mm (± 1/4 in)

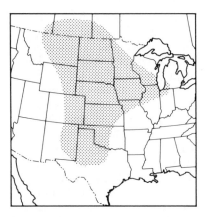

long, each with a tuft of feathery bristles.

HABITAT

Prairies and native pastures.

PARTS USED

Roots (harvested in fall or winter)—cooked.

FOOD USE

Gayfeather is an extremely drought-resistant, perennial, prairie plant with a small edible root that varies greatly in palatability. The Kiowa gathered the roots in the spring when they were sweet and baked them over a fire. Later in the season, they reported "the roots have a greasy taste and consequently are not gathered or eaten" (Vestal and Schultes, 1939, p. 61). Gayfeather was considered by the Kiowa to be one of the "ancient" foods, and because of its ability to withstand drought, was still an important part of their diet in the 1930s (ibid., p. 72). Although widely distributed over the prairies, gayfeather is seldom mentioned as a food source of native people. Melvin Gilmore (1931b, p. 102) listed two species of gayfeather, *L. pycnostachya* Michx. and *L. squarrosa* (L.) Michx., among the vegetal remains of the Ozark Bluff-dweller culture, but whether they were used as food, medicine, or ornament could not be determined.

During the month of August, the Osage harvested corn, squash, wild plums, and the roots of gayfeather (Matthews, 1961, p. 478). These roots would be stored in caches for use during the winter and probably became sweeter with age, as their stored starch changed to sugar. The Tewa, who live in New Mexico, were also reported to have eaten the roots of *L. punctata* (Robbins et al., 1916, p. 57).

The Indians of the prairies were observant of nature because their lives depended upon it. They understood the growth cycles of many plants and animals and how these cycles related to the seasons. When gayfeather's flower spikes became purple blazing stars, the people of the northern Plains tribes would say: "Now the Arikara corn is coming into condition for eating. Let us go and visit them" (Gilmore, 1926, p. 14). Then they would journey to the Arikara villages, bringing with them handicrafts and gifts from the products of the natural resources of their own country. While enjoying feasts of green corn with their Arikara hosts, they would trade and exchange commodities (ibid.).

When I dug up the thick-spiked gayfeather, *L. pycnostachya*, in November on a northeast Kansas prairie, I found the egg-sized root, which was several years old, to be quite woody, bitter, and medicinal-tasting. However, a year-old segment of the larger root was tender, somewhat starchy, and sweet. Roots of *L. punctata*, which I harvested in late December in central Kansas, were woody and tough—there was no edible part to them. Julian Steyermark (1981, p. 1474) described this species as "carrot-flavored." In the *Flora of the Great Plains* (Great Plains Flora Assn., 1986, p. 973), *L. mucronata* DC. is considered a distinct species from *L. punctata* because of its globose corm, which may be too woody to eat. Because of the extreme variation in roots within and between species and the lack of knowledge about their medicinal use, gayfeather roots cannot be recommended as a food.

The reason that gayfeather is considered an ancient food source is the remarkable ability of this plant to withstand drought. Dotted gayfeather is the only species to form a long taproot, with lateral roots that reach out at various levels for water. John Ernest Weaver (1968, p. 83), who was the foremost prairie ecologist of his time, reported from one of his University of Nebraska experiments that during the first summer of its growth, gayfeather developed a deep taproot that was out of all proportion to its shoot. "In August, plants only 5 inches tall, and with only 2 leaves, possessed taproots 33 to 38 inches deep which had accumulated some reserve food."

Prairie vegetation has the ability to withstand the extremes of nature, in particular an extremely unpredictable and at times almost nonexistent supply of water. Finding edible plants on the prairie during a time of extreme drought is a difficult task and would require a thorough knowledge of plants that could serve as marginal food sources. Gayfeather is one plant that was found during the extreme drought of the 1930s, as J. H. Robertson reported (1939, p. 460):

Four days spent on this prairie in the middle of July, 1936, served to impress one with the severity of physical and biotic factors so extreme that even native vegetation could not endure them uninjured. Daily maximum temperatures ranging from 104 to 111 degrees F. accompanied by relative humidities of 19 to 24 percent, continued strong winds, glaring sunlight, and subnormal precipitation combined to make the grasses crackle underfoot like wheat stubble. Only the very deeply rooted false boneset,

Kuhnia eupatorioides, and blazing star, Liatris punctata, *appeared unhampered by drought, and they were borne down and partly eaten by hordes of grasshoppers. Amorpha folded its leaves during the day and opened them only slightly by sunrise although the nights were cool and relatively calm. Foliage of all other plants was partly or entirely dead.*

CULTIVATION

The numerous purple flower spikes of gayfeather add a beautiful accent to the prairie's subtle colors. These flower spikes can be cut, dried in a dark place, and used for winter bouquets. The flowers will continue to open as they dry, so for greatest vividness of color they should be picked before they are in full bloom.

Gayfeather can be grown in a sunny location, from either seeds or root-cuttings planted in late fall or early spring. Seeds are very small—about 139,000 to a pound—and only about three seeds are planted per square foot when a large area is being restored to prairie because they have a high germination rate. In one experiment 96 percent germination rate was obtained after seeds were stratified and then germinated at 79°F (Salac et al., 1978, p. 4).

Lomatium foeniculaceum
Prairie Parsley

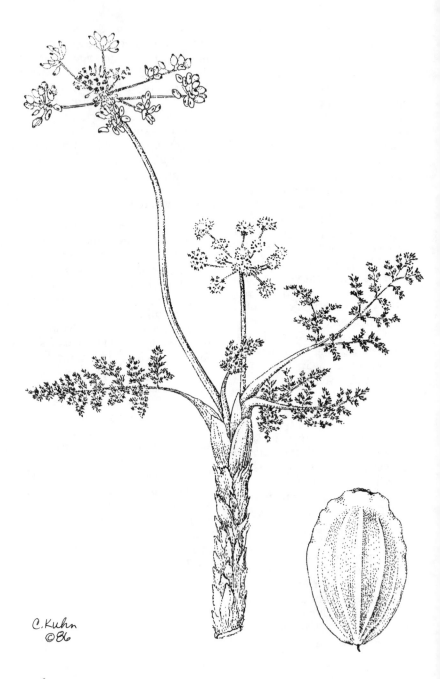

C. Kuhn
©86

COMMON NAMES

Prairie parsley and carrot-leafed lomatium. (The entire family is often referred to as biscuit root because of the biscuit flour that was obtained from some of the more starchy rooted species.)

INDIAN NAMES

The closely related *Lomatium orientale* Coult. & Rose, found on drier soils, was named "sahijela tatinpsinla" (Cheyenne turnip) by the Lakota (Munson, 1981, p. 237).

SCIENTIFIC NAME

Loma'tium foeniculac'eum (Nutt.) Coult. & Rose is a member of the Apiaceae (Parsley Family). *Lomatium* means "small border," an allusion to the wings of the fruits. The species name, *foeniculaceum*, means "like fennel" because of the resemblance to that plant.

DESCRIPTION

Perennial herbs 1–5 dm (4–20 in) tall, growing from thickened taproots, no stems. Leaves clustered at base, oval to oblong, 1–20 cm (⅜–8 in) long, divided into 3 segments, each of these dissected into narrow segments. Flowers in round, flat clusters on stalks rising above leaves, from Mar to May; petals 5, separate, small, yellow. Fruits dry, oval to oblong, 5–12 mm (³⁄₁₆–½ in) long, with longitudinal ribs and wings.

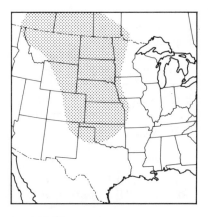

HABITAT

Prairie hillsides and rocky limestone soils.

PARTS USED

Leaves (spring)—salad; roots of some species (fall)—for flour.

FOOD USE

All species of *Lomatium* are edible. They are best known from the northwestern United States, where the large starchy roots were often dried, ground into flour, and made into biscuits or cakes by several different Indian tribes. Prairie parsley is one of the first prairie plants to bloom in the spring. Its edible leaves have a strong parsley taste and are good in salads. The roots are thin, woody, and taste slightly sweet (like purple poppy mallow roots), but with a bitter aftertaste.

The Lakota were reported to use the roots of the closely related wild parsley, *L. orientale*, for food

(Munson, 1981, p. 237). The other biscuit root of the Prairie Bioregion, *L. macrocarpum* (H. & A.) Coult. & Rose, is found only in the northern part of the region (North Dakota, Montana, Saskatchewan, and Manitoba). It was a food of the Flathead (Hart, 1976, p. 26).

CULTIVATION

In *Hortus Third—A Concise Dictionary of Plants Cultivated in* the United States and Canada (Bailey, 1976, p. 678), biscuit roots are listed as "sometimes planted in wild gardens." Stratified seeds can be planted in the spring or they can be propagated by root cuttings in the spring or fall. Their yellow clusters of flowers, blooming early in the spring, are a nice addition to a rock garden.

Monarda fistulosa
Beebalm

C.Kuhn
© 86

Beebalm, wild bergamot, horse-mint, American horsemint, Os-wego tea, plains bee balm, and fern mint.

INDIAN NAMES

The Dakota call it "hehaka ta pe-zhuta" (elk medicine) or "hehaka to wote" (food of the elk) (Gil-more, 1977, p. 59). They also iden-tify a second form or variety that they call "wahpe washtemna" (fragrant leaves). The Omaha and Ponca call it "pezhe pa" (bitter herb) and refer to the second va-riety as "izna-kithe-ige," which they used as a fragrant pomade for the hair (ibid.). The Pawnee recog-nize four varieties of beebalm. The lowest form they call "tsu-sahtu" (ill-smelling); the second is "tsostu" (no translation given); the third is "tsakus tawirat" (shot many times still fighting); and the fourth, which is their most desir-able, is "parakaha" (fragrant) (ibid.). The Blackfoot call it "ma-ne-ka-pe" (young man) (Johnston, 1970, p. 319). The Kiowa call it "po-et-on-sai-on" (perfume plant) (Vestal and Schultes, 1939, p. 49).

SCIENTIFIC NAME

Monar'da fistulo'sa L. is a mem-ber of the Lamiaceae (Mint Fam-ily). *Monarda* is named after Nicholas Monardes, a sixteenth-century physician of Seville, who wrote about medicinal and other useful plants of the New World.

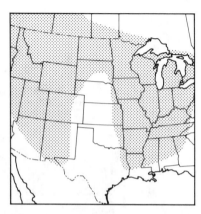

The species name *fistulosa* means "tubular," in reference to the shape of the flowers.

DESCRIPTION

Perennial herbs from creeping rhi-zomes, 3–12 dm (12–48 in) tall; stems square, usually hairy above, sometimes branched. Leaves op-posite, oval to lance-shaped, 3–10 cm (1¼–4 in) long, lower surfaces hairy, margins toothed or nearly entire, fragrant. Flowers in round clusters at ends of branches from Jun to Sep; petals fused into tubes, separating into two sections, up-per 1 erect, lower 1 bent back-wards, 2–3.5 cm (¾–1⅜ in) long, pale to dark lavender, rarely white. Fruits dry, hard, 1.5–2 mm (±¹⁄₁₆ in) long, brownish or blackish.

HABITAT

Prairie hillsides, pastures, road-sides, banks, and occasionally in open woods, usually in rocky soil.

Leaves (spring)—boiled for tea, seasoning, chewed raw, or dried.

Beebalm is a fragrant herb used for seasoning, tea, perfume, and medicine. Most of the uses of beebalm recorded in the Southwest probably apply also to the Prairie Bioregion, because the culture groups of these areas had contact with each other.

The Apache, Tewa, and Hopi all used beebalm as a seasoning (Opler, 1936, p. 57). The Tewa used it to flavor meat (Robbins et al., 1916, p. 57). The Hopi also ate lemon mint or lemon beebalm, *M. citriodora* Cerv. (whose range also extends into the Prairie Bioregion), which "was boiled and only eaten with hares" (Fewkes, 1896, p. 19). This same plant was used by the Tewa at Hano as a cooked green, eaten by itself (Castetter, 1935, p. 34). The Pueblo people of the Southwest also used beebalm to season beans and stews and dried it for winter use. In addition, they chewed the spotted beebalm, *M. pectinata* Nutt., while hunting (ibid.).

Beebalm, like most plants in the Mint Family, produces a flavorful and healthful tea with medicinal properties. The Lakota make a tea from beebalm that is soothing for sore throats; they use it for other medicinal purposes as well. The plant has a sweet fragrance and a distinctive taste and is often chewed while people are singing and dancing (Rogers, 1980a, p. 78; Munson, 1981, p. 236). According to J. Owen Dorsey (1894, p. 454), a specific variety of beebalm is used during the Sun dance ceremony. The Flathead sprinkled the pulverized leaves on meat that was being preserved, to keep insects away (Hart, 1976, p. 71). The Omaha and Ponca use the finer, more delicate of their two named varieties of beebalm as a fragrance for a pomade for the hair (Gilmore, 1977, p. 59). The Dakota also identified two varieties of beebalm and the Pawnee identified four, which indicate their knowledge, observation, and use of this plant. Melvin Gilmore (1977, p. 60) planted the two Dakota-named varieties of beebalm and observed them for five years. He verified that there were distinct differences between them, though they are scientifically classified as the same species.

Beebalm varies greatly in taste, depending on the variety and the time of year it is harvested. In my experience, the early leaves are the best substitute for mint tea, the later leaves (until the plants flower) make the best oregano-tasting seasoning, and the old leaves are the hottest (they can be used in a hot sauce). The flowers are an attractive edible garnish in a salad. The fragrance and taste of beebalm comes from the compounds limonene, carvacrol, and cymene. Lemon mint, *M. punctata* L., is a rich source of thymol, which is fungicidal, anthelmintic and a major ingredient in antiseptic preparations (Foster, 1984, p. 67).

A handsome ornamental, beebalm is an easy plant to grow in the garden. The Hopi reportedly cultivated this plant for its greens, and dried them for winter use (Krochmal, 1954, p. 13). Melvin Gilmore (1977, p. 7) believed that beebalm was inadvertently propagated by the Indians of the Missouri River region, who used it widely and spread its seeds while carrying plants after harvesting them.

Many varieties of the red-flowered beebalm, *M. didyma* L., have been selected for their colorful flowers. These and our native beebalm are easy to propagate by division of the roots in spring, or the tiny seeds can be planted in spring. They should be sown one-fourth inch deep. Beebalm is attractive to bees and they feed on the flowers quite heavily.

Opuntia macrorhiza
Prickly Pear

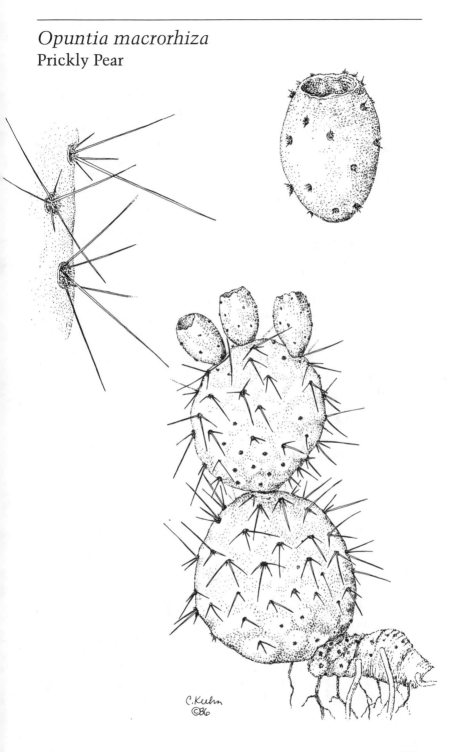

C. Kuhn
©86

COMMON NAMES

Prickly pear, bigroot prickly pear, prickly pear cactus, beavertail, and tuna.

INDIAN NAMES

The Cheyenne name is "mah-ta'-o-munst" (prickly fruit) (Grinnell, 1962, p. 180). The Dakota names are "unhce'la blaska'" (flat cactus) and "unhce'la tan'ka" (large cactus) (Rogers, 1980a, p. 61). The Blackfoot name for the closely related prickly pear, *Opuntia polyacantha* Haw., is "ohkotowatisis" (many sharp points) (Johnston, 1970, p. 316). The Cree call the edible fruits of prickly pears "meyicimina" (feces berry) (Mandelbaum, 1940, p. 203).

SCIENTIFIC NAME

Opun'tia macrorhi'za Engelm. is a member of the Cactaceae (Cactus Family). *Opuntia* comes from an old name for a plant whose identity is unknown. The species name, *macrorhiza*, means "large root" because the main roots are often enlarged and tuberous.

DESCRIPTION

Perennials, low-growing in clumps, less than 12 cm (4¾ in) tall; stems bluish green with flattened, roundish segments (pads), to 10 cm (4 in) long, bearing clusters of tiny spines with 1–6 larger spines to 5 cm (2 in) long. Leaves cylindrical, fleshy, falling off read-

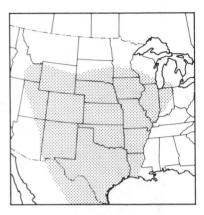

ily. Flowers solitary near margins of older stem segments, usually yellow, sometimes reddish. Fruits fleshy, narrowly egg-shaped, to 5 cm (2 in) long, purple to red, bearing clusters of tiny spines; seeds disk-shaped, with broad rims.

HABITAT

Dry soils of prairie, pastures, roadsides, and rocky glades.

PARTS USED

Ripe fruit, pads, buds, and flowers (summer)—raw, cooked, or dried; flower buds despined, roasted, dried and cooked in stews; seeds (fall)—dried, then roasted and ground for soup.

FOOD USE

There are several species of prickly pear, all of which have spines, large, attractive flowers, and edible fleshy parts. They thrive under dry conditions.

The Cheyenne ate the fruits of prickly pear raw and dried. Before the occupancy of the southern part of the Prairie Bioregion by white people, the gathering, drying, and storing of the fruit of the cactus was one of the Southern Cheyenne women's important duties. George Bird Grinnell reported how this was done (1962, p. 180):

The fruit was collected in parfleche sacks and was then put on the ground in little piles and stirred and swept over by small brushes, made of twigs of sagebrush, until most of the thorns had been removed. The women, having first made little finger tips of deerskin to protect the ends of the fingers, then went over the piles and removed the last thorns from the fruit. When this had been done, the fruit was split, the seeds removed and thrown away, and the flesh dried in the sun. This was used to stew with meat and game, and made a gelatinous thickening to the soup. This fruit was still gathered as above as late as 1915.

Cheyenne men also used prickly pears as a source of water when on war parties (Hooper, 1975, in Hart, 1981, p. 17).

The Comanche (Carlson and Jones, 1939, p. 527) and Pawnee (Palmer, 1871, p. 418) dried the unripe fruit to be stored and cooked later with meat and other foods. Some Indian tribes (probably of the Southwest, but perhaps including the Pawnee) boiled the despined, unripe fruit in water for 10 to 12 hours, until it had the consistency of applesauce. After being allowed to ferment a little, this substance was reported by Edward Palmer to be "stimulating and nutritious" (ibid.).

The summer of 1856 was remembered by the Kiowa as "Se'nalo' K-a'do'" (prickly-pear sun dance). They gave this distinctive name because they held the sun dance "at a place where there was an abundance of prickly-pears, at the mouth of a small creek, probably Caddo or Rate creek, entering the Arkansas river about 10 miles below Bent's Fort, near La Junta, Colorado. It was held late in the fall, when the prickly-pears were ripe, instead of in midsummer, as usual, and the women gathered a large quantity" (Mooney, 1895, p. 301).

Lieutenant J. W. Abert, returning via a circuitous route from Bent's Fort to St. Louis, arrived at a place along the Canadian River near what is today the New Mexico–Texas border on September 2, 1845, and reported that there was a great abundance of prickly pear in fruit. "In their flavor the raspberry and water melon seem mingled. . . . We frequently paid dear for handling them, the little spines being barbed like a fish hook" (McKelvey, 1955, p. 935). And nine days later in the Texas Panhandle, he reported: "The party was disappointed more than once, by finding that Indians had already gathered the plums, grapes and cactus fruits of the region" (ibid.).

Although prickly pear fruits are tasty, they must be eaten carefully. I once did not remove a ⅛-inch-long spine, which ended up in my lip and spoiled the pleasure of eating the fruit. Also, there is the belief among several tribes of Indians in the Southwest (including the Pima, Yuma, and Apache) that eating too many tunas (as the fruits are called) will give the eater chills (Niethammer, 1974, p. 15).

Prickly pear stems are often mentioned as an emergency food. Melvin Gilmore (1913b, p. 366) reported that when food was scarce, the Dakota would first clear the spines from the stems of prickly pear and then roast them for food. After the Battle of Beecher Island along the Republican River in northeast Colorado, while General Forsythe and his troops were stranded on that sandy island until September 25, 1868, prickly pear was one of their few foods. Chauncey B. Whitney was a scout for the campaign and he reported the situation (1911, p. 296):

September 22, 1868—No Indians to-day; Killed a coyote this morning, which was very good. Most of the horse meat gone. Found some prickly pears, which were very good. Are looking anx-

iously for succor (relief) from the fort.

September 23, 1868—Still looking anxiously for relief. Starvation is staring us in the face; nothing but horse meat.

The taste of raw prickly pear pads is similar to that of raw okra or cucumber. Early settlers from Montana reported that the pads can be prepared by boiling, which loosens the skin, so that it and the prickles can be easily removed; then the soft, pulpy interior can be fried for an excellent dish (Blankenship, 1905, p. 17).

Prickly pear pads and seeds were major food sources for Indians living 6,000 years ago in Hinds Cave in the Lower Pecos areas of southwest Texas. Remains of prickly pear pads were frequently seen in their coprolites, 74 percent of which also contained fragments of broken prickly pear seeds. These seeds were darkened in color, suggesting that the fruit had been heat-treated before being pounded or ground and then eaten (Williams-Dean, 1978, p. 193).

Also, prickly pear pollen was found in large quantities in some coprolites, indicating that the flowers were a seasonal food source. In her experiments relating to the coprolites at this site, Glenna Williams-Dean discovered that the petals can be picked from the flowers of prickly pears and that the sexual parts can be left in place to be fertilized and develop into the fruit. She concluded that it is possible to ingest prickly pear pollen while eating the petals, without necessarily endangering the production of the edible fruit (ibid., p. 152).

CULTIVATION

Prickly pears are unusual food sources, whose odd growth forms, ability to survive in dry habitats, and showy flowers make them a good plant to cultivate where their spines will not interfere with human activities. Father de Smet, a Catholic missionary, who made extensive travels across the prairies, deserts, and mountains of the West, described the prickly pear in a letter he wrote in 1842:

The flowers of these are beautiful, and known among Botanists by the name of Cactus Americana. *They have already been naturalized in the flower gardens of Europe. The colors of the handsomest roses are less pure and lively than the carnation of this beautiful flower. The exterior of the chalice is adorned with all the shades of red and green. The petals are evasated like those of the lily. It is better adapted than the rose to serve as an emblem of the vain pleasures of this nether world, for the thorns that surround it are more numerous (Thwaites, 1906, 27: 255).*

The Indian fig, O. *ficus-indica* (L.) Mill., is a very large cactus grown in Mexico and other areas for its edible fruits. The fruits and pads can be found among the other exotic produce in supermarkets, packaged with instructions for preparation. Luther Burbank developed Burbank's Spineless Cactus, a form completely without spines, from this species. It was extensively promoted for a few years as a forage plant for arid and semi-arid regions. Richard Felger lists O. *phaeacantha* Engelm., which is native to the southwestern prairies, as one of the Sonoran Desert food plants that has potential as an agricultural crop (Felger, 1979, p. 11). On a garden or field scale, prickly pears are easy to propagate because their stems and pads will root if covered with moist soil. They also can be grown from seed.

Oxalis violacea
Violet Wood Sorrel

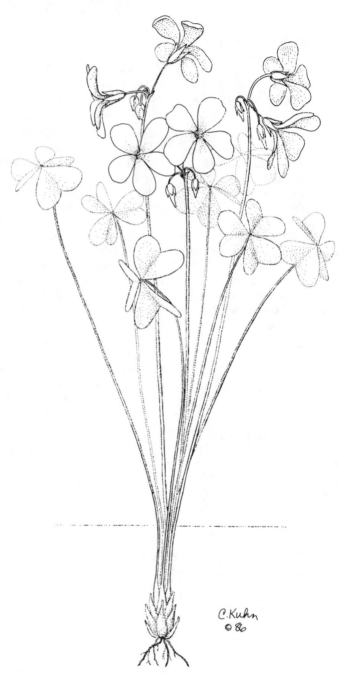

C.Kuhn
© 86

Violet wood sorrel, wood sorrel, oxalis, and sheep sorrel. (Sorrel generally refers to a plant with sour juice.)

INDIAN NAMES

The Omaha and Ponca call the wood sorrels "hade-sathe" (sour herb) (Gilmore, 1977, p. 46). The Pawnee call them "skidadihorit," a name having reference to its taste, which they describe as "sour like salt"; they also call them "askirawiyu" (foot, water, stands), possibly referring to the plant's being found near some wet areas; and "kait" (salt), probably in reference to its taste (ibid.). The Kiowa name for yellow wood sorrel (*Oxalis stricta* L.), "aw-tawt-an-ya," also means "salt weed" (Vestal and Schultes, 1939, p. 35).

SCIENTIFIC NAME

Ox'alis viola'cea L. is a member of the Oxalidaceae (Wood Sorrel Family). *Oxalis* is derived from the Greek word meaning "sour." The species name *violacea* means "violet in color."

DESCRIPTION

Perennial herbs, growing from scaly bulbs, no stems. Leaves clustered at base, 6–12 cm (2.5–4¾ in) long, divided into 3 leaflets, heart-shaped. Flowers in round flat-topped clusters at ends of stalks taller than leaves, from Apr to Jun (sometimes again in Sep and Oct);

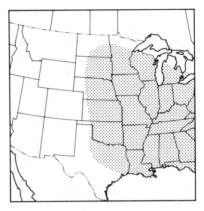

petals 5, separate, 1–2 cm (⅜–¾ in) long, pinkish-purple, rarely white. Fruits dry, cylindrical, 4–6 mm (± ¼ in) long, opening longitudinally to release seeds with netlike pattern on surfaces.

Oxalis stricta differs in having a stem, leaves often purplish, and yellow petals.

HABITAT

For violet wood sorrel—moist prairies, rocky open woods, thickets, and waste ground. Yellow wood sorrel is less common in prairies, but also grows in fields and woods and along roadsides.

PARTS USED

Leaves, flowers, bulbs (spring or fall)—raw or cooked. (Raw leaves should be eaten in moderation because they contain oxalates.)

FOOD USE

Violet wood sorrel and yellow wood sorrel are eaten for their sour, salty flavor. Both plants are enjoyed by children. Yellow wood sorrel was probably the first wild edible plant that I learned about as a child. Slightly older and more knowledgeable neighborhood kids taught me that its leaves and seed pods were edible. The pods were called "little bananas." Pawnee children were also very fond of sorrels, especially the violet wood sorrel, eating the leaves, flowers, flower stems, and bulbs (Gilmore, 1977, p. 46). The Pawnee observed that buffalo ate yellow wood sorrel (ibid.).

Because of the acid nature of the plant, the Kiowa chewed the leaves to relieve thirst when on long walks and when perspiring freely. Their name, which means "salt weed", may "indicate that there was an early realization that the loss of salt through perspiration may be counteracted by chewing the leaves of this plant" (Vestal and Schultes, 1939, pp. 35, 36). Other reports of the use of wood sorrel by the native people of the prairies note that the Osage ate the leaves (Matthews, 1961, p. 454) and that the Omaha used the flowers medicinally as a poultice for swellings (Gilmore, 1913a, p. 335).

Wood sorrel is nutritious, con-taining significant amounts of vitamin A in its leaves (Arnason et al., 1981, pp. 2238–2239). The flowers make an attractive and tasty garnish for salads. However, it also has a large quantity of oxa-lates, which produce the charac-teristic acid, tart taste. Oxalates tie up calcium in the body and in large quantities can cause poison-ing. One symptom of too much oxalate is painful or swollen taste buds, which I got one spring from eating too much raw Swiss chard. A little bit of wood sorrel in a salad or as a snack is healthful; a large quantity in one meal or eaten over a period of time can be harmful.

CULTIVATION

Oca is a species of wood sorrel, *O. tuberosa* Mol., which is cultivated in the high Andes for its edible tubers. Our wood sorrels also have edible tubers, but they are not large enough to suggest cultivat-ing them. Violet wood sorrel is a nice addition to a rock garden or wildflower garden. Seeds are hard to find, but plants can be propa-gated by transplanting them or di-viding their bulbs in the spring. Wood sorrel spreads by runners after it becomes established.

Physalis heterophylla
Ground Cherry

C. Kuhn
©86

COMMON NAMES

Ground cherry, clammy ground cherry, husk tomato, tomatillo, Chinese lantern, strawberry tomato, and popweed.

INDIAN NAMES

The Omaha and Ponca name "pegatush" and the Pawnee name "nikaktspak" both translate as "forehead, to pop" in reference to children using the inflated husk to pop on their foreheads (Gilmore, 1977, p. 61). The Lakota name is "tamni'ohpi hu" (womb, fetal membrane, and nest). Perhaps this is in reference to the fetuslike fruit inside the persistent husk (Rogers, 1980a, p. 96).

SCIENTIFIC NAME

Phy'salis heterophyl'la Nees. is a member of the Solanaceae (Nightshade Family). *Physalis* means "plant with a bladdery husk" and the species name *heterophylla* means "having leaves of different forms," in reference to the variations in leaf margins.

DESCRIPTION

Perennial herbs with erect, hairy stems 1.5–5 cm (⅝–2 in) tall, sometimes much branched. Leaves alternate, egg-shaped to 4-angled, 5–10 cm (2–4 in) long, both surfaces hairy, margins irregularly wavy or toothed. Flowers solitary among leaves, nodding, from May to Oct; each one 1.5–2.5 cm (⅝–1 in) wide, petals fused

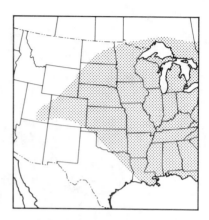

into bell-shaped tubes, with 5 shallow lobes at tops, yellow with 5 dark spots at bases of tubes. Fruits fleshy, round, yellowish, enclosed in papery, inflated coverings, egg-shaped and 3–4 cm (1¼–1⅝ in) long.

HABITAT

Prairies, pastures, roadsides, fields, open woods, and waste ground.

PARTS USED

Ripe fruit (late summer, fall)—raw or cooked. (Note that green fruit and other parts of the plant may be poisonous.)

FOOD USE

The ripe fruit of the ground cherry tastes somewhat like a tomato and was a widely used food source of the Indians. Elias Yanovsky, a USDA chemist, in *Food Plants of the North American Indians* (1936, p. 56) lists 10 species of ground cherries that were used for food. (This book still contains the most comprehensive catalogue of native plant foods in North America.)

Buffalobird Woman informed Gilbert Wilson (1916, unpublished notes, in Nickel, 1974, p. 69) that ground cherries were scarce on the Hidatsa reservation in North Dakota. However, when they were found, they were collected and eaten fresh in the field or brought back to the lodges. When occasionally found in quantity, the fruits were pounded and shaped into patties similar to those made with chokecherries.

Melvin Gilmore reported (1977, p. 61) that the fruits of *P. heterophylla* were made into a sauce for food by the tribes of the Missouri River region. The Zuni (in New Mexico) made a highly prized sauce by boiling ground cherries and then grinding them in a mortar with raw onions, chili peppers, and coriander seeds (Stevenson, 1915, p. 70). The Dakota also used ground cherries as a food source. When they first saw figs, they called them "white man's ground cherries" (Gilmore, 1977, p. 61).

As a child, I learned that popping the papery husk surrounding the fruit was even more fun than popping a paper bag. Kiowa children also enjoyed "pop weed." Their mothers often gathered the fruits of *P. lobata* Torr., and in the 1930s they reportedly used them to make jelly (Vestal and Schultes, 1939, p. 50).

Carbonized seeds of ground cherry are present in archaeological sites across the Prairie Bioregion. They were found at the Mitchell site (along Firesteel Creek, near present-day Mitchell, South Dakota), which was occupied from 985–1125 A.D. (Benn, 1974, p. 236). At Phillips Spring, along the Pomme de Terre River in western Missouri, carbonized seed remains were found among cultivated gourds from 2280 B.C. and squash from 313 B.C. (King, 1980, pp. 217, 231). Carbonized seeds of ground cherries were also found at the Anasazi pueblos at Salmon Ruin in northwest New Mexico (Doebley, 1981, p. 181). All of these findings indicate the widespread importance of this plant as an ancient food source.

Dr. H. H. Rusby (1906, p. 448) noted: "The best of the common species is the clammy ground cherry (*Physalis heterophylla*), . . . The yellow berry is enclosed in an inflated, veiny, cone-shaped husk or hull, which is the accrescent calyx. The fruit has a peculiar strong flavor, not entirely agreeable before full maturity, but then giving place almost altogether to an agreeable sweetness."

Only the ripe fruit of the ground cherry is edible. Green fruits, leaves, and roots may be poisonous. Because *Physalis* is a member of the Nightshade Family, confusion with poisonous relatives should be avoided. Ground cherries are nutritious, but contain only average amounts of vitamins and minerals when compared to domesticated fruits and vegetables (Watt and Merrill, 1963, p. 133). However, their unusual taste, when added to a sauce or eaten by themselves, can make a meal or snack more enjoyable.

CULTIVATION

There are several varieties of ground cherries (those called tomatillos are a Mexican species) that have been cultivated for food or for their novel and attractive flowers and fruits. Matilda Stevenson, an ethnologist, reported (1915, p. 70) that she observed Zuni women cultivating ground cherries, *P. longifolia* Nutt., in their small gardens.

Liberty Hyde Bailey, an eminent horticulturalist from Cornell University, said the following about ground cherries: "The plant is worthy of a place in every home garden" (Medsger, 1976, p. 80). Seeds of the annual species can be germinated indoors or sown directly outside where there is a long growing season. The perennial species, such as the clammy ground cherry, can easily be started by division, or from a root cutting or seed.

Proboscidea louisianica
Devil's Claw

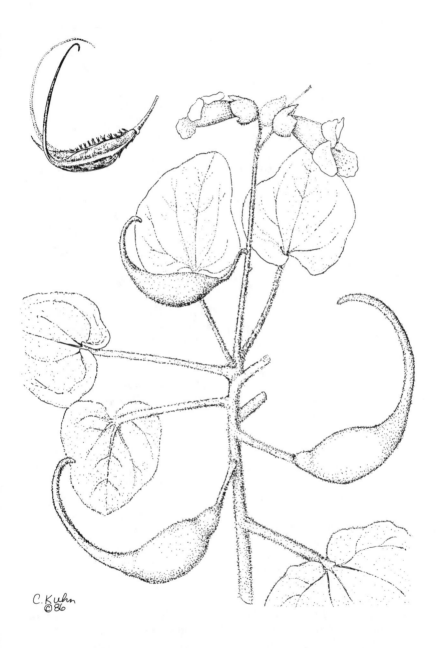

C. Kuhn
© 86

Devil's claw, unicorn plant, unicorn flower, proboscis flower, double claw, cat's claw, and ram's horn. (The common names are related to the curiously shaped fruit, which when green is horn- or proboscis-shaped. When mature and dry, the fruit splits apart, forming two sharply hooked horns or claws, which are adapted for the dispersal of the seed by clasping the legs of animals.)

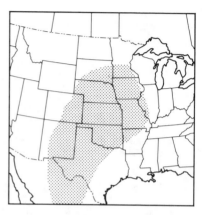

INDIAN NAMES

None were found in the sources consulted.

SCIENTIFIC NAME

Proboscid'ea louisian'ica (P. Mill.) Thell. is a member of the Pedaliaceae (Unicorn-plant Family). *Proboscidea* comes from the Greek word for the proboscis-shaped fruit. The species name, *louisianica*, means "of Louisiana."

DESCRIPTION

Annual herbs with strong, horsy odor, 1.5–6 dm (6–24 in) tall, erect or lying on ground, covered with glandular hairs. Leaves opposite, occasionally alternate near tops of plants, kidney-shaped or rounded, 3–20 cm (1¼–8 in) long, margins wavy or entire. Flowers in elongated groups at ends of branches, extending above leaves, from Jun to Oct; petals 5, fused into tubes 2–3.5 cm (¾–1⅜ in) long, swollen on one side, opening into broad lobes at tops, 1.5–2 cm (⅝–¾ in) long, whitish to purplish, mottled with yellow or reddish purple, inner tubes with conspicuous reddish-purple spots. Fruits cylindrical, with fleshy coverings that fall away, leaving dry structures to 1 dm (4 in) long, each with a curved beak, longer than the body, which splits into 2 parts; body of fruit opening into 2 sections; seeds dull black, narrowly oval, with corky texture.

HABITAT

Fields, pastures, and waste ground (usually in sandy soil).

PARTS USED

Immature green fruits (summer)—cooked or pickled; seeds (late summer, fall)—raw or cooked.

FOOD USE

This unusual plant, with its claw-shaped seed capsules, strong musky odor, and beautiful large whitish flowers, is not an important prairie plant, with distribution primarily in the drier, southern area. Its native habitat was probably limited to disturbed areas such as flood plains, stream banks, buffalo wallows, and prairie dog towns. The extensive use of this plant in the Southwest by over 30 native culture groups for basketry fibers, food, and other purposes (Nabhan, 1981, p. 139), indicates its probable use within the Prairie Bioregion, although no records exist.

Alfred Whiting, an ethnobotanist, reported (1939, p. 92) that the Hopi believed that the long spines of devil's claw drew lightning and hence rain, so the plant was never weeded out of fields. Also in the Southwest, devil's claw was grown for its fibers, which were used in basketry, "although in certain cultures (e.g., the Papago), seeds were no doubt eaten also" (Nabhan et al., 1981, p. 140).

Edward Palmer was one of the first observers to write about the use of devil's claw by native people. In 1871 (p. 422), he reported: "The Apache Indians gather the half mature seed-pods of this plant (*Proboscidea fragrans*), and cook them with various other substances. The pods, when ripe, are armed with two sharp, horn-like projections, and, being softened and split open, are used on braided work to ornament willow baskets."

The Chiricahua Apache, when interviewed in the 1930s, told of an instance of devil's claw being used as a food source in the Prairie Bioregion under unusual circumstances. They reported that the seeds (probably of *P. louisianica*) were eaten by Indian boys when they were prisoners of war in Oklahoma (Opler, 1936, p. 45).

The following two historical descriptions point out the ecological niche that devil's claw filled on the prairies as a colonizer of disturbed habitats. After leaving the Rocky Mountains in July of 1820, Dr. E. James, the botanist of the Long Expedition, reported near present-day Pueblo, Colorado: "The barren cedar ridges, are succeeded by still more desolate plains, with scarce a green, or living thing upon them, except here and there a tuft of grass . . . among the few stinted and withered grasses. . . . Near the river and in spots of uncommon fertility, the unicorn plant was growing in considerable perfection" (McKelvey, 1955, p. 227).

Lieutenant J. W. Abert, while in the panhandle of Texas in 1845, reported the devil's claw to be "overgrowing the dry ponds or buffalo wallows" (ibid., p. 936). Buffalo wallows were created when buffalo tore up the soil with their hooves and rolled in the dust. These circular depressions, 10 to 30 feet across, are still visible in pastures and some rangeland of the region.

To prepare devil's claw for eating, first pick pods during the summer when they are still young and tender, and only an inch or two long. Then wash them under running water, brushing with a vegetable brush to remove as many hairs as possible, and boil them in salted water until tender (Niethammer, 1974, p. 95). At this stage of growth, devil's claw fruits also make good pickles.

The nutritive value of the seeds was determined in the 1950s, when there was considerable interest in using this plant as an oil seed crop. An analysis of devil's claw growing in the southwestern United States showed that the seeds normally range between 35 and 43 percent oil and 20 and 35 percent protein (Nabhan et al., 1981, p. 1954). It was reported that "this semi-drying oil resembles cottonseed oil and sunflower oil, and could be satisfactorily substituted for cottonseed oil in the manufacture of salad oils and shortenings" (Krochmal et al., 1954, p. 8).

CULTIVATION

Devil's claw was domesticated in the Southwest. A white-seeded va-

riety of the closely related species, *P. parviflora*, was developed in the Southwest, and it and other species were cultivated by the Pima, Papago, and numerous other Indian tribes (Nabhan et al., 1981, p. 140). The antiquity of the cultivation of this plant is suggested by the anthropologist Richard Yarnell, who believes that it must have taken several centuries to selectively breed devil's claw plants with longer pods and white seeds (Ford, 1981, p. 21).

Devil's claw was mentioned as early as 1841 in William Kenrick's *New American Orchardist* as a garden crop grown for food in the eastern United States (Hedrick, 1919, p. 356). It is a potential oil seed crop for semiarid regions and an unusual and attractive ornamental. Devil's claw is an annual that can be planted from seed in the spring after the ground has warmed, or, in areas with shorter growing seasons, it can be started inside and transplanted.

Prunus americana
Wild Plum

C. Kuhn
©86

COMMON NAMES

Wild plum, American plum, sand-hill plum, Osage plum, river plum, sand cherry, thorn plum, wild yellow plum, red plum, August plum, goose plum, hog plum, and sloe.

INDIAN NAMES

The Kiowa name is "pank-ai-da-lo" (sour plum or thick-rind plum) (Vestal and Schultes, 1939, p. 29). The following names are for "plum" and "plum tree" respectively: Dakota, "kante" and "kante-hu"; Omaha and Ponca, "kande" and "kande-hi"; Winnebago, "kantsh" and "kantsh-hu"; and Pawnee, "niwaharit" and "ni-waharit-nahaapi" (Gilmore, 1977, p. 35). The Cheyenne name for the wild plum is "mak-u-mins'" (great-berry) (Grinnell, 1962, p. 177). The Comanche differentiate between "yuseke" the "early plum," "parawaseke" the "late summer plum," and "kusiseke" the "fall plum" (although it is possible that one of these may have been the wild cherry, *Prunus serotina* Ehrh., or chokecherry, *P. virginiana* L.) (Carlson and Jones, 1939, p. 523). The Dakota name for the closely related sand cherry, *P. pumila* L. var. *besseyi* (Bailey) Gl., is "aonyeyapi" (Gilmore, 1977, p. 36) and the Cheyenne name is "muh'-ko-ta-mins" (Grinnell, 1962, p. 177). Translations of these names indicate that the fruit is properly picked when the hu-

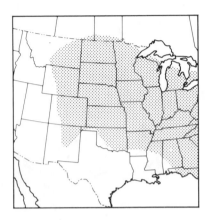

man scent of the picker is carried away from the plant, because it could give a bad taste to the fruit.

SCIENTIFIC NAME

Pru'nus american'a Marsh. is a member of the Rosaceae (Rose Family). *Prunus* comes from the Greek "prounos," an ancient name for the plum tree. The species name *americana* means "American."

DESCRIPTION

Shrubs or small trees, 3–8 m (3–24 ft) tall, usually forming thickets, small branches sometimes spiny. Leaves alternate, egg-shaped to oval, 6–10 cm (2⅜–4 in) long, upper surfaces shiny green, lower surfaces slightly hairy, margins sharply toothed. Flowers in groups of 2–5 at ends of branchlets, usually appearing before leaves in Apr and May; petals 5, separate, oval, 8–12 mm (⁵⁄₁₆–½ in) long, white.

Fruits fleshy, oval, 2–2.7 cm (¾–1¹⁄₁₆ in) long, reddish–purple or yellowish, each containing 1 seed.

HABITAT

Thickets in prairies, woodlands, pastures, and along roadsides and riverbanks.

PARTS USED

Ripe fruit (midsummer to fall)—raw, cooked, in jelly, or dried.

FOOD USE

Wild plum fruit was extensively consumed by the Indians of the prairies, either fresh or made into a sauce. Plums were also pitted and dried, although the Pawnee reportedly often dried them without removing the pits (Gilmore, 1977, p. 35).

The Kiowa were also reported to use large quantities of the wild plum, *P. gracilis* Engelm. & Gray, which they dried whole for winter use or pounded up and made into cakes (Vestal and Schultes, 1939, p. 30). Sometimes the Comanche obtained wild plums in the winter by tracking pack rats to their nests and taking the supply the rats had hoarded (Carlson and Jones, 1939, p. 526). The Omaha planted their corn, beans, and squash when the wild plum came into bloom, and the Teton Dakota used the sprouts or young growth of the wild plum as a wand in the "waunyanpi" ceremony (Gilmore, 1977, p. 35).

The Lakota call the moon that corresponds to our month of August, "Kan'tas'a wi," which means "red plum moon," because this is the time when the much-appreciated wild plums are ripe (Rogers, 1980a, p. 90).

The Comanche ate the fresh fruits of the Chickasaw plum, *P. angustifolia* Marsh., or they pitted and dried them for winter use (Carlson and Jones, 1939, p. 526). The Crow (Blankenship, 1905, p. 19), Assiniboin (Denig, 1930, p. 583), and Kiowa (Vestal and Schultes, 1939, p. 30) used the wild plum similarly.

In the monumental work, *The Cheyenne Indians* (1962, p. 177), George Bird Grinnell discusses their use of plants for food and medicine. About the wild plum, *P. americana*, he reported the following:

Near the camps the wild plums seldom ripen because the children pick and eat them green; but in places at a distance many plums were gathered by women, and were stoned, dried in the sun, and kept for winter. When ripe plums thus dried were cooked by boiling, they became almost like cooked fresh plums. They were a great and rather unusual delicacy. The plum bushes do not always produce a crop, and sometimes for several successive seasons, as a result of late frosts or from some other cause, no crop of plums is had.

At Walth Bay, a late prehistoric archaeological site near present-day Mobridge, South Dakota, plum pits were found that were very similar to those of *P. americana* (Nickel, 1974, p. 40). Plum pits identified from this same species were recovered from the Dodd and Phillips Ranch sites from the Oahe Dam area in Stanley County, South Dakota (Lehmer, 1954, p. 163). Also, pits of this species were identified from the Hill site, near Guide Rock, Nebraska, which was a historic Pawnee Indian village (Wedel, 1936, p. 59).

The early explorers and travelers of the Prairie Bioregion often mentioned wild plums in their journals and diaries and also appreciated them as food. On July 11, 1804, Captain William Clark of the Lewis and Clark expedition reported seeing the Osage (wild) plum near the "clear water" of the Nemaha River in what is now southeast Nebraska. He provided the following description of this location from the top of an Indian burial mound on a hill overlooking the Nemaha River valley, about two miles above that river's confluence with the Missouri River:

I had an extensive view of the Serounding Plains, which afforded one of the most pleasing prospect ever beheld, under me a Butifull River of Clear Water of about 80 yards wide Meandering thro: a leavel and extensive

meadow, as far as I could See, the prospect much enlivened by the fiew Trees and Srubs which is bordering the bank of the river, and the Creeks & runs falling into it, The bottom land is covered with Grass of about 4 and ½ feet high, and appears as leavel as a smoth surfice, the 2nd bottom is also covered with Grass and rich weeds & flours, interspersed with copses of Osage Plumb (Thwaites, 1904, 1: 75).

George Catlin reported in 1837 (1973, 2:52), while near the Red River in present-day southern Oklahoma:

The next hour we would be trailing through broad and verdant valleys of green prairies, into which we had descended; and often-times find our progress completely arrested by hundreds of acres of small plum-trees of four or six feet in height; so closely woven and interlocked together, as entirely to dispute our progress, and sending us several miles around; when every bush that was in sight was so loaded with the weight of its delicious wild fruit, that they were in many instances literally without leaves on their branches, and bent quite to the ground.

Edwin James, the botanist for the Long expedition, reported on August 13, 1819, while in the area that is now the Texas panhandle, during heat of 105 degrees in the shade, that the grapes were "ripe and delicious" and that the "Osage plum" was beginning to ripen (McKelvey, 1955, p. 231). He wrote four days later, from near the Antelope Hills, in present-day western Oklahoma: "The grapes and plums, so abundant in this portion of the country, are eaten by turkies and black bears, and the plums by wolves or jackals, as we conclude, from observing plumstones in the excrement of one of those animals" (Thwaites, 1905, 16: 135). A few weeks later, near the confluence of the Verdigris and the Arkansas rivers, near present-day Muskogee, Oklahoma, James reported: "Late in the afternoon, we struck the Osage trace, leading from their village to the trading establishment, at the confluence of the Verdigrise, whither we now direct our course. Our evening encampment was at a small ravine, in which were some plum bushes, bearing fruit, yet unripe, of a fine red colour, and, without the slightest exaggeration, as closely situated on many of the branches as onions when tied on ropes of straw for exportation" (ibid., p. 282).

Wild plums were appreciated by other military expeditions. Lieutenant J. W. Abert noted on September 4, 1845, again near the Antelope Hills, that after crossing a sandy, waterless waste, their sufferings were "greatly alleviated by the refreshing fruit of the plum tree . . . equal to any of the cultivated varieties" (McKelvey, 1955,

p. 936). Captain L. C. Easton reported on August 23, 1849, while near Elm Creek (also called Muddy Creek) near what is now Arapahoe, Nebraska: "We crossed a stream to day on which there was a number of Elm Trees—Saw Three Elk to day—Passed a Grove of Plum Trees, from which our party gathered large quantities of the finest wild fruit I ever saw" (Mattes, 1952, p. 406).

The early settlers of Kansas and other states in the Prairie Bioregion made extensive use of wild plums. William Chase Stevens, a University of Kansas botanist, reported (1961, p. 257) that "*Prunus angustifolia watsoni*, sand Chickasaw plum, named in honor of Dr. Louis Watson of Ellis, Kansas, is a dwarf edition of the species, 3–10 feet high, prevailing in the sandy soil of the Republican, Saline, and Arkansas River valleys, where it often occurs in large societies, affording great quantities of desirable, though acidly austere fruit, which the early settlers gathered by the bushel and wagon-load, using it for sauce, pies, puddings, jelly, and preserves."

The sand cherry, *P. pumila* var. *besseyi*, is another species that has a dwarf form. It loves sandy areas and is quite common in the Nebraska Sand Hills. It is used like a plum, although its fruit, which varies in quality, is more like a tart cherry in shape and taste. The Dakota believed that if "a person gathering cherries moves in the direction contrary to the wind the cherries will be good and sweet, but on the other hand if he moves with the wind the cherries will be bitter and astringent" (Gilmore, 1977, p. 36).

Both wild plum fruits and seeds have economic value. One study of an unnamed *Prunus* species indicated that the seeds contain 33.4 percent protein and 44.3 percent oil by dry weight (Earle and Jones, 1962, p. 229). Like chokecherry pits, these seeds contain hydrocyanic (prussic) acid, which is poisonous. However, this acid can be removed by modern processing or the traditional method of cooking or pounding and then drying.

CULTIVATION

Wild plums should be encouraged or cultivated for their fragrant spring blossoms; for their sweet, yet tart, wild-tasting fruit; and for their ability to withstand drought and insects. The Native Americans may not have actually cultivated wild plums, but they certainly encouraged them. Dr. V. Havard of the U.S. Army reported in "Food Plants of the North American Indians (1895, p. 103):

Prunus Americana Marsh, and P. nigra Ait., our two species of Wild Yellow or Red Plum, were, according to several authorities, planted by the New England and Canadian natives, and from the many forms discovered further west it is not improbable that this culture extended to the Mississippi. Some forty-five horticultural kinds derived from them are described by Prof. Bailey, and it is not assuming too much to suppose that several of them are due to variations initiated by Indian industry. It is probable enough, however, that the native orchard was seldom regularly planted, but oftener the accidental result of seeds dropped in the vicinity of camping grounds and villages.

Melvin Gilmore (1931a, p. 92) was able to learn from the Pawnee why they had plum thickets around their original home sites on their reservation land in Oklahoma:

The four tribes of the Pawnee confederacy were removed by the United States federal government from their own home country in Nebraska to Oklahoma in 1873, because the Pawnee land in Nebraska was desired by white people. In the places where they were temporarily settled for the first year of their residence in Oklahoma there may now be seen many thickets of Prunus americana, the wild plum, which abounds in the region of the Pawnee country in Nebraska. . . . Pawnees have told me that the plum thickets which now appear in the places where they lived in Oklahoma, when they first came from Nebraska, are the results of the seeds which they threw out after using the dried plums which they had brought in their food provision from Nebraska.

The encouragement of wild plums by prairie settlers is another factor in their widespread distribution. On the farm where I grew up, near Guide Rock, Nebraska, there were several plum thickets that were established by my grandfather in the 1930s for erosion control. At least one of these thickets is still thriving, and it is possible that the local wild colonies of plums, from which these thickets were started, were brought to this area by the Pawnee, who previously had a village nearby.

In 1979 Marty Bender and I were research associates at the Land Institute of Salina, Kansas. On September 23, while on a botany field trip, we stopped a few miles northwest of Abilene, Kansas, in some sand hills where we saw an abundance of the Watson variety of wild plum. These plums were growing in a dense colony and were loaded with fruit (like those James described on the Long expedition), and we decided to harvest some to see how much they would yield. Marty later dried, pitted, and weighed the plums and calculated that if the yield was extrapolated, it would amount to 3,820 pounds of pitted dried plums per acre. In addition, there would be 2,660 pounds of dried kernels per acre (which could be processed and used as a protein source) (Bender, 1984, personal communication).

I have noticed that colonies of wild plums seem to lose their productiveness with size and age, and disease is common in large colonies. Sometimes colonies die in the center, while the edges keep expanding. For these reasons, such extrapolations of yields of wild plums may not be valid; but they are undoubtedly a highly productive wild fruit.

Wild plums have been recommended for their drought resistance and widely planted in shelter belts in the western Great Plains. They also make good wildlife habitat and are effective in erosion control because their roots hold the soil. Their thorny branches catch tumbleweeds, leaves, and other plant materials, which, when wind storms occur during times of drought, provide an effective means of slowing wind erosion of soil. The following cultivated varieties of plum are direct descendents of our wild species, *P. americana:* Blackhawk, Hawkeye, and De Soto. The Watson variety, mentioned earlier, is a named cultivar of *P. angustifolia.* Wild plums can be planted from seed and they are rather easy to transplant.

Prunus virginiana
Chokecherry

C. Kuhn
© 86

Chokecherry, common chokecherry, wild cherry, and chokeberry. (The fruit is very astringent
or puckery, which accounts for
the "choke" part of its name.)

INDIAN NAMES

The Omaha and Ponca name is
"nonpa-zhinga" (little cherry)
(Gilmore, 1977, p. 36). The Pawnee name is "nahaapi nakaaruts"
(cherry tree) (ibid.). The Lakota
name is "canpa'-hu" (bitter-wood
stem) (Rogers, 1980b, p. 57). The
Blackfoot names are "pukkeep"
(choke cherry) and "puck-keep"
(the berry) (Johnston, 1970, pp.
313, 314). The Cheyenne name is
"monotse" (berries) (Hart, 1981, p.
35). No translations were given for
the following names, but most of
them probably would translate as
"chokecherry." The Kiowa name
is "o-hpan-ai-gaw" (Vestal and
Schultes, 1939, p. 30). The Crow
name is "malupwa" and the Flathead name is "schla scha" (Blankenship, 1905, p. 19). The Assiniboin name is "cham-pah" (Denig,
1930, p. 583). The Osage name is
"goonpa" (Munson, 1981, p. 237).

SCIENTIFIC NAME

Prun'us virginia'na L. is a member of the Rosaceae (Rose Family).
Prunus is from the Greek "prounos," an ancient name for the
plum tree. The species name *virginiana* means "of Virginia."

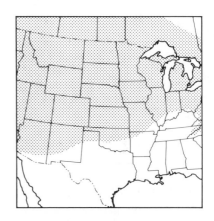

DESCRIPTION

Shrubs or small trees, 2–6 m (6–
20 ft) tall, small branches red-
brown to dark brown, often forming thickets. Leaves alternate, egg-
shaped to broadly elliptic, 4–12
cm (1⅝–4¾ in) long, upper surfaces dark green and lustrous,
lower surfaces grayish green, usually hairy along veins, margins
toothed. Flowers in dense, elongated groups at ends of branches,
in Apr and May; petals 5, separate,
rounded, 3–4 mm (⅛–3/16 in) long,
white. Fruits fleshy, round, 8–11
mm (5/16–7/16 in) in diam, red or
bluish purple to black, each containing 1 seed.

HABITAT

Rich soils, thickets, fence rows,
roadsides, borders of woods, sandy
and rocky soil on hillsides and ravine banks.

Ripe fruit (June to October)—
cooked, or dried; kernels—
pounded with fruit and dried
(pemmican); green sticks as skew-
ers to flavor meat while cooking;
bark for tea.

FOOD USE

The chokecherry is one of the
most widely distributed woody
species in North America. It has
edible, medicinal, and poisonous
properties. Its dried fruit was a
staple food of the Indians and the
trappers and traders of the prairies
and Rocky Mountains.

Food is more than just something to eat. It is a focus for social
interaction and an important part
of the local or tribal economy,
thus bringing wealth and health to
those who provide it. It also is
often part of the relationship to
the supernatural and a statement
of cultural identity. The chokecherry, although unnoticed by
most people in the Prairie Bioregion today, once was a focal
point of all these interactions and
relationships. Also, as much as
any food, it brought together the
roles of male buffalo hunter and
female berry gatherer, as foods
made by the women from chokecherries were usually combined
with buffalo meat and fat.

The chokecherry was esteemed
by all the tribes of the area and
was one of the principal ingredi-
ents in the dried meat/fat/fruit

mixture called pemmican. Chokecherries were staples of the Blackfoot, gathered by the women, and
divided

*into those to be preserved whole
and those to be crushed. The ones
for storage were greased and dried
in the sun, then put away in the
fawn-skin bags. The others were
crushed on a stone and mixed
with backfat for pemmican or
added to soups, etc. The juice
from the crushing was given as a
special drink to husbands or the
favorite child.*

*Choke-cherry sticks were gath-
ered, stripped of their bark and
inserted into roasting meat for
spice (Hellson, 1974, p. 104).*

Although the Blackfoot had no
cereal from which breads could be
made, pemmican provided a similar dietary staple, richer and
higher in protein.

*For this, the best cuts of buffalo
were dried in the usual manner.
Then they were pounded on a
stone until fine. . . . Just before
pounding, the pieces of dried
meat were held over the fire to
make them soft and oily. Marrow
and other fats were heated and
mixed with the pounded meats,
after which crushed wild cherries
were worked into the mass.
Often, a few leaves from the peppermint plant were added in order to give flavor to it. The whole
was then packed into parfleche or
other bags, a compact sticky*

mass, easily preserved and good for eating without further preparation. While the flesh of buffalo was preferred for pemmican that of deer and elk would be used if at hand (Wissler, 1910, p. 22).

Gilbert Wilson reported the following Hidatsa accounts of crushed chokecherries being made into patties or balls. "In contrast to her use of plums, Buffalobird Woman reported the storage for winter of one or two bushels of dried chokecherry patties" (personal note in Nickel, 1974, p. 71). Owl-woman demonstrated the traditional method of preparing chokecherry balls:

A stone pestle and mortar were set on a piece of skin. The cherries, not quite ripe, were crushed three or four at a time, pits and pulp together. The resultant mass was brushed onto the skin. When a sufficient quantity had been mashed in this way, the cherries were patted between the palms of both hands into the form of a ball. Holding the ball in her left hand and covering it with her right hand, Owl-woman squeezed the chokecherry mass between the thumb and forefinger of her left hand and simultaneously drew her hands backward to deposit the cherries in an elongated lump on the board provided for the purpose. They were left there to dry before eating (Weitzner, 1979, pp. 215–216).

The Crow (ibid.) and the Cheyenne (Grinnell, 1962, p. 178) were also reported to have made chokecherry patties. Chokecherries were the most abundant and important fruits of the Cheyenne; they also made a fine berry pemmican from the fruits, pits, and kernels (ibid.). The pits were probably included because they were difficult to remove and the nutritious kernels they contained added more flavor.

Captain William Clark of the Lewis and Clark expedition reported on December 23, 1804, on the Mandan whose village (in present-day North Dakota) was near the fort built by the expedition party: "Great numbers of indians of all discriptions Came to the fort many of them bringing Corn to trade, the *little Crow,* loaded his wife & Sun with Corn for us, Cap. Lewis gave him a few presents as also his wife, She made a kittle of boiled Cimmins, beens, Corn & Choke Cherries with the Stones, which was palitable This Desh is Considered, as a treat among those people" (Thwaites, 1904, 1: 240).

The Arikara dried chokecherries, but to obtain an even larger supply, they traded with the Dakota. Chokecherries were valuable, and the Arikara traded two measures of their own shelled corn for every measure of the Dakotas' dried chokecherries (Gilmore, 1926, p. 15).

The harvest of chokecherries

was of great importance to the domestic economy of the Dakota. They even named the month of July, in which cherries ripen in their area, "Chanpa-sapa-wi," which literally means "Black-cherry moon" (Gilmore, 1913b, p. 365). During this month, "the people travel for miles to the streams where the cherries are abundant and there go into camp and work up the cherries while they last, or until they have prepared as great a quantity as they require" (ibid.). Also, the sun dance began on the day of the full moon when the cherries were ripe (Gilmore, 1977, p. 37).

The Omaha and other corn-growing tribes cooked dried chokecherry cakes with their corn in the winter. Melvin Gilmore (1926b, p. 14) described the preparation of a chokecherry/corn meal mush: "Another variety of mush was made entirely from meal of parched corn. For this dish flint corn was parched and then ground in a mortar. The resulting meal was then made into mush by boiling in the usual way by gradually dropping and stirring into boiling water. This mush was then served either plain or seasoned with suet, or with dried chokecherries mixed in during the cooking process. . . . For flavoring the mush the cakes of dried cherries were broken up and stirred into the meal."

Other tribes of the prairies that harvested and ate chokecherries were the Osage (Munson, 1981, p. 237), Kiowa (Vestal and Schultes,

1939, p. 31), Pawnee (Gilmore, 1977, p. 36), Assiniboin (Kennedy, 1961, p. xlii), and the Comanche (Wallace and Hoebel, 1972, p. 74). One other food use, which may have occurred in the prairies as well was reported from the Great Lakes Bioregion, where the Meskakis used the bark of chokecherry to make a beverage (Smith, 1928, p. 283).

Chokecherries were an ancient food of the Prairie Indians. Chokecherry seeds have been found in archaeological remains at a historic Pawnee village site (the Hill site) near what is now Guide Rock, Nebraska (Wedel, 1936, p. 59), and at 11 of 62 sites in North and South Dakota (Nickel, 1977, p. 55). Only corn, beans, and squash were reported to have been found at a greater number of these archaeological sites.

Chokecherries, in the form of pemmican, were an important food source for the early trappers, traders, and explorers of the region. Edwin James, botanist for the Long Expedition, mentioned in 1820 the chokecherry cakes that they purchased in an area now in Oklahoma: "The squaw had in her possession a quantity of small flat blackish cakes, which on tasting we found very palatable. Having purchased some of them, we ascertained that they were composed of the wild cherry, of which both pulp and stone were pounded together, until the latter is broken into fragments, then mixed with grease, and dried in the sun" (Thwaites, 1905, 16: 218). John Charles Frémont, an early explorer of the West for the United States government, reported on July 29, 1842, while along the Oregon Trail near the Red Buttes (about 15 miles southwest of Casper, Wyoming), that on the banks of the river "the cherries are not yet ripe, but in the thickets were numerous fresh tracks of the grizzly bear, which are very fond of this fruit" (Jackson and Spence, 1970, p. 243).

The common name "chokecherry" is a warning of the effect caused by eating the astringent raw fruit. One early description of the chokeberry comes from William Wood in 1746 (Erichsen-Brown, 1979, p. 153). "The cherrie trees yeeld great store of cheries which grow in clusters like grapes; they be much smaller than our English Cherrie, nothing neere so good; if they be not very ripe, they so furre the mouth that the tongue will cleave to the roofe, and the throate wax horse with swallowing those red Bullies (as I may call them) being little better in taste." Captain Clark of the Lewis and Clark expedition also reported while in the Rocky Mountains on August 24, 1805 that we have "nothing to eate but Choke Cherries & red haws, which act in different ways So as to make us Sick" (Thwaites, 1905, 3: 33).

Chokecherries contain hydrocyanic (prussic) acid in the leaves and also in their pits. Prussic acid

poisoning from chokecherries has been reported in livestock when the leaves were eaten (Phillips, 1959, section 4, p. 20) and in small children from eating fruits (Harrington, 1967, p. 257). The cyanide is often associated with a bitter taste and seems to vary considerably between plants. Indians of California ate large quantities of the pits of the closely related wild cherry, *P. ilicifolia* (Nutt.) Walp., but only after proper preparation through leaching, boiling, or drying (Timbrook, 1982, pp. 166–170).

The native methods of preserving chokecherries throughout the prairies were to dry them thoroughly or cook them, which served to break down the prussic acid contained in the pits. Chokecherry pits have not been analyzed for their nutritional content, but one study indicated that the seeds of an unnamed *Prunus* species contain 33.4 percent protein and 43.3 percent oil by dry weight (Earle and Jones, 1962, p. 229). Drying chokecherries not only preserves them and reduces the prussic acid, but it also seems to improve their flavor by sweetening them (Harrington, 1957, p. 257).

Chokecherries have been used by early settlers and many wild foods enthusiasts for sauces, jelly, and wine. In "Native Economic Plants of Montana" (Blankenship, 1905, p. 20), chokecherry marmalade and chokecherry butter are described: "The latter is usually prepared by cooking the ripe fruit, straining out the seeds and skins through a colander, and then mixing with an equal quantity of plums or crab-apples to modify the harsh, astringent, bitterish taste. This 'choke-cherry butter' thus prepared is highly prized as a food throughout the state and only needs to be better known to become a regular article of commerce."

CULTIVATION

Chokecherries were not reported to have been cultivated by the Indians of the prairies, but they were probably encouraged and may have been planted. The Cree, who presently live in the subarctic of Ontario, have increased local populations of chokecherries by planting them around their homes (Black, 1978, p. 260). It is quite likely that chokecherries may have been similarly planted in the vicinity of Indian village sites on the prairies.

Chokecherries have frequently been planted for their attractive spring flowers and to provide food and habitat for wildlife. Although chokecherries do have some disease and insect problems, they should be considered more often for landscaping because of their drought resistance. For this reason, they have been planted in shelterbelts in the Great Plains. The largest crop of chokecherries I have ever seen was in a shelterbelt planting in central Nebraska—the bushes were loaded for the entire one-half mile planting.

A yellow-fruited cultivar of the chokecherry called "Xanthocarpa" is available through nurseries. Chokecherries do best in a sunny location with a rich soil and can be transplanted from the wild in early spring.

Psoralea esculenta
Prairie Turnip

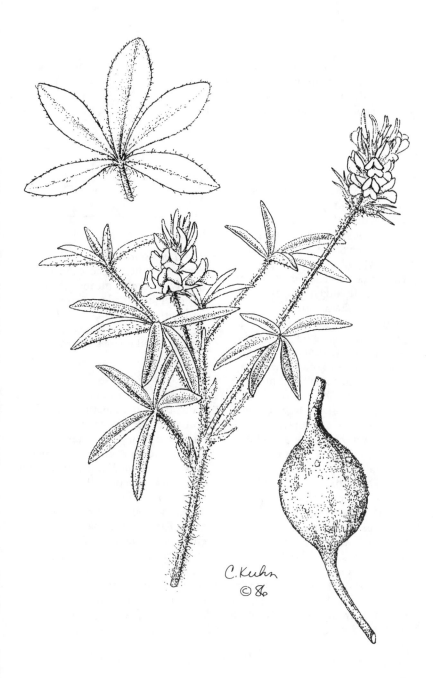

C. Kuhn
© 86

COMMON NAMES

Prairie wild turnip, Indian turnip, Dakota turnip, Indian breadroot, prairie potato, pomme blanche, pomme de prairie, prairie apple, white apple, ground apple, and tipsin.

INDIAN NAMES

The Osage name is "dogoe" of which "do" translates as "potato" (Munson 1981, p. 237). The Blackfoot name is "mas" or "mats" (elk food) (Johnston, 1970, p. 314). The Winnebago name is "tdokewihi" (hungry) (Gilmore, 1977, p. 40). The Crow name is "aha" or "esharusha" (Blankenship, 1905, p. 20); the Arikara name is "hsu' proka" (Gilmore, 1926b, p. 14); the Omaha and Ponca name is "nug'the" (Gilmore, 1977, p. 40); and the Dakota name is "tipsin" and "tipsinna" (Gilmore, 1977, p. 40).

Dr. J. R. Walker suggested that the Dakota name "tipsinna" is derived from their name for wild rice, "psin". Also, "tinta" is the Dakota word for "prairie," and "na" is a suffix diminutive. The compound word "tinta-psin-na" could be translated as "the little wild rice of the prairie." The Dakota previously inhabited a region nearer the Great Lakes where wild rice was more abundant. Perhaps because the prairie turnip became a major staple food, replacing wild rice in their diet, it was given this name (ibid., p. 4).

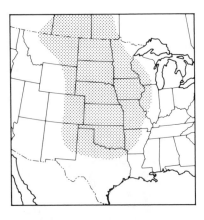

SCIENTIFIC NAME

Psoral'ea esculen'ta Pursh is a member of the Fabaceae (Bean Family). *Psoralea* means "scabby," and refers to the plant being covered with glandular dots. The species name *esculenta* means "edible," and refers to its root.

DESCRIPTION

Perennial herbs from deep taproots, thick and tapering toward both ends; stems 1–3, densely hairy, 5–15 cm (2–6 in) tall. Leaves alternate, often appearing clustered, divided into 5 fingerlike segments, each one elliptic to egg-shaped, 2.5–5 cm (1–2 in) long, dotted with glands, lower surfaces hairy. Flowers in dense, narrow clusters among leaves from May to Jul; petals 5, upper 1 larger and erect, 2 lower ones boat-shaped, 2 wings at sides, blue fading to yellowish. Fruits dry, oval, 5–7 mm (³⁄₁₆–⁵⁄₁₆ in) long, each tapering into a beak; seeds oblong, plump, olive green and often purple-spotted or dark brown.

HABITAT

Undisturbed prairies, hay meadows, and well-managed pastures, scattered across dry and rocky (often limestone) soils.

PARTS USED

Roots (harvested from late May to July, after flowers blossom, but before leaves and stem dry, break off, and blow away)—fresh, cooked, or dried.

FOOD USE

The prairie turnip was probably the most important wild food gathered by Indians who lived on the prairies. The general impression given by eyewitnesses and ethnographers is that prairie turnip roots were dug whenever they were encountered, and that other activities were often suspended until an adequate supply was accumulated for the present and future (Reid, 1977, p. 323).

One of the earliest descriptions of the use of the prairie turnip comes from the Lewis and Clark expedition. On May 8, 1805, near the confluence of the Milk and Missouri rivers in northern Montana, Captain Meriwether Lewis observed:

This root forms a considerable article of food with the Indians of the Missouri, who for this purpose prepare them in several ways. They are esteemed good at all seasons of the year . . . (and) are sought and gathered by the provident part of the natives for their winter store, when collected they are striped of their rhind and strong on small throngs or chords and exposed to the sun or plased in the smoke of their fires to dry; when well dryed they will keep for several years, provided they are not permitted to become moist or damp (Thwaites, 1904, 2: 11).

On July 17, 1858, Henry Youle Hind, near Qu'Appelle Mission in southeastern Saskatchewan, found a party of Plains Cree collecting prairie turnip and reported: "Many bushels had been collected by the squaws and children and when we came to their tents were employed in peeling the roots, cutting them into shreds and drying them in the sun. I saw many roots as large as the egg of a goose, and . . . some of even larger dimensions" (Mandelbaum, 1940, p. 202).

Young Indian children helped gather wild foods. Lakota mothers told their children that prairie turnips "point to each other" (Gilmore, 1929, p. 190–191). When the children had noted in which way the branches were pointing, they were sent in that general direction to find the next plant. This saved the mothers from searching for plants, kept the children happily busy, and made a game of their work (ibid.).

Prairie turnips were so important that they even influenced the tribes' selection of hunting grounds. For the Omaha, "the line of march taken on the tribal buffalo hunt was sometimes determined by the localities where this desirable plant grew in abundance" (Fletcher and La Flesche, 1911, p. 341).

In all these instances, it should be pointed out that the women were the gatherers of prairie turnip roots (as was true with most edible prairie plants) and that their work was considered of great importance to the tribe. Perhaps explorers and travelers (nearly all of whom were men) would have paid more attention to the prairie turnip, if its procurement and processing had involved men.

The prairie turnip was the main commodity the Dakota traded to the Arikara for corn. Whole prairie turnip roots were traded by the string, which was made by braiding the stringy roots of the prairie turnip together like onion or garlic tops. A standard length of a braided string of roots was one arm-reach. Split and dried roots were traded by the burden basket measure which has a capacity about equal to a bushel. A fair trade was one burden basket of shelled corn for four strings of prairie turnip roots and one burden basket of dried, split roots. In comparison, one burden basket full of shelled corn was considered equal to one good buffalo robe (Gilmore, 1926, p. 14).

Few remains of the prairie turnip are found in archeological excavations. This is not surprising because fleshy tubers do not preserve well and the only process that would have lead to carbonization (and preservation) would have been roasting, which was not the major method of preparation (Nickel, 1974, p. 24).

The use of the prairie turnip since antiquity is indicated by its appearance in the mythology of the Omaha, Lakota, and other tribes. John Neihardt in *When the Tree Flowered* (1951, p. 156) retells a Lakota myth, "Falling Star, the Savior," in which a pregnant woman digs prairie turnips so vigorously that she digs completely through the earth and falls into the prairie sky.

Its importance since ancient times is further illustrated by the fact that Indians used the prairie turnip in their geographical place names. Wild Turnip Hill, near Lethbridge, Alberta, was called "Mas'-etomo" by the Blackfoot, which directly translated means "Turnip Butte." And the area in the vicinity of Cowley, Alberta, was known to the Blackfoot as "Akai-sowkaas," which meant "Many Prairie Turnips" (Johnston, 1970, p. 314).

The season to harvest prairie turnip roots is quite short. It may be only six or seven weeks from the time the plants bloom until the above ground portions dry and blow away. Roots I have dug before the plant flowers in the spring were woody and there was little stored starchy material. Harvest time starts in May in the southern prairies and continues through July in the northern prairies.

Much energy can be spent harvesting prairie turnips. In fact, I have sometimes just quit digging when the soil was so hard or gravelly that it required a pickax. It seems amazing that the roots are able to penetrate such hard ground. A shovel or digging tool is necessary to harvest the hen-to-goose-egg-sized roots, which are often four or more inches below the surface. Indians made digging sticks of ash or other wood and hardened the points with fire. Tips of elk antlers were also used. My favorite natural digging tool is a sharp deer antler tine.

The prairie turnip has been reported to be both bland and nourishing and it has significant quantities of starch and protein (Kaldy et al., 1980, p. 355). To my sense of taste, the raw root tastes like a mild turnip with a little bit of a beany flavor. Dr. V. Havard of the United States Army, reported (1895, p. 108) that "raw it has a very palatable farinaceous flavor entirely devoid of bitterness . . . and may be found to this day in all tents of the Sioux Indians for whom it has always been a staple food. They generally eat it cooked, and as they appreciate the advantages of a mixed 'pot-au-feu,' boil it with tripe, fattened pup or venison."

Meriwether Lewis observed, in the summer of 1805, the manner in which prairie turnips were prepared for eating. In describing whole dried turnips, he wrote:

They usually pound them between two stones placed on a piece of parchment, untill they reduce it to a fine powder, thus prepared they thicken their soupe with it; sometimes they also boil these dryed roots with their meat without breaking them; when green they are generally boiled with their meat, sometimes mashing them or otherwise as they think proper. They also prepare an agreeable dish with them by boiling and mashing them and adding the marrow grease of the buffaloe and some buries, until the whole be of the consistency of hasty pudding. They also eat this root roasted and frequently make hearty meals of it without sustaining any inconvenience or injury there from" (Thwaites, 1904, 2: 11).

Among the Plains Cree, Hind ate a "sort of pudding made of the flour of the (prairie turnip) root and the mesaskatomina berry [service berry, *Amelanchier canadensis* (L.) Medic.] which is very palatable and a favorite dish" (Mandelbaum, 1940, p. 202). In 1833, feasting with the Blackfoot at Fort McKenzie (located on the Missouri River at the mouth of the Marias River in north-central Montana), Prince Maximilian of Wied reported that "a wooden dish was set before each of us, containing boiled beaver's tail, with prairie turnips, (the) pomme blanche. The beaver's tail was cut into small slices, and was boiled very tender. It did not taste amiss, and is reckoned a good dish even in the United States" (Thwaites, 1906, 23: 133).

Humans are not alone in appreciating the prairie turnip. It was noted in several early reports, including that of Dr. F. V. Hayden (1862, p. 369), that the root was a favorite of the now extinct Plains grizzly bear.

The prairie turnip's ecological niche is undisturbed prairie, but it is not a dominant species. Although it has little forage value (Forest Service, 1937, p. W157), prairie turnip decreases under heavy grazing by livestock. It is seldom found in disturbed areas, does not spread quickly, and, therefore, is not very common. This once abundant prairie plant has been a victim of the tremendous change in the economy of the Prairie Bioregion, which transformed the native prairies into cropland and pastures. I was surprised to see it on my walk across Kansas, in the four-feet-deep remnant ruts of the Santa Fe Trail west of Wilsey, Kansas, because this is a disturbed habitat. Before harvesting this plant, be certain

there are others nearby to maintain a future healthy population.

Two related species, *P. cuspidata* Pursh and *P. hypogaea* Nutt., also have edible roots, though smaller, and are found on short-grass prairies within the western range of *P. esculenta.*

CULTIVATION

It is doubtful that Plains Indians ever directly cultivated the prairie turnip, and there is no ethnographic evidence to support this idea. The seedlings take two to four years to produce mature taproots, which is slow-growing for a crop. It has been suggested by Kenneth Reid in an article in *Plains Anthropologist* (1977, p. 325) that because of its importance and abundance, the prairie turnip may have been reseeded by Plains gatherers. They may have replanted it during harvest, because the ripe seeds in the dried top could be planted in the hole where the root had grown. The resulting "turnip patches," harvested and seeded on a staggered schedule from year to year, could have played a significant role in the prehistoric definition of gathering territories.

Cultivation of the prairie turnip was attempted in the early 1800s by a French naturalist, Lamarea Picot. Philander Prescott, Superintendent of Farming for the Sioux, reported (1849, p. 452) that Mr. Picot "has lately incurred a very considerable expense to obtain the seed, which he has carried to France, believing that it is capable of cultivation, and may form a substitute both for potato and wheat." Prairie turnip roots, called "picquotiane," were grown and marketed for a while in France, but apparently the experiment was not a success (Maisch, 1889, p. 346).

Prairie turnips can be transplanted successfully to a sunny location. Seeds that have been scarified by filing a notch through the hard outer hull will sprout when planted in the spring (Reichart, 1983, p. 110).

Rhus glabra
Smooth Sumac

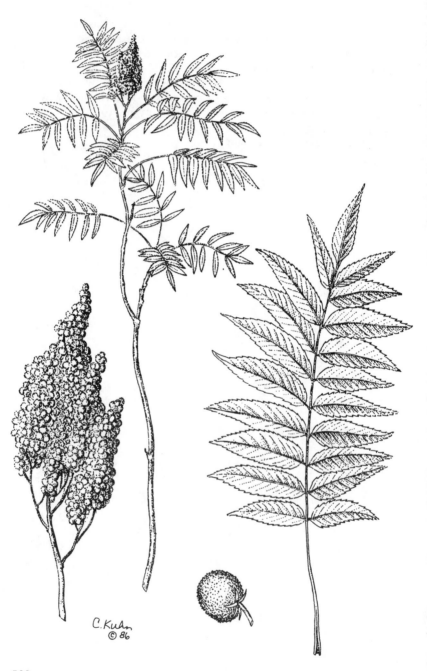

C. Kuhn
© 86

COMMON NAMES

Smooth sumac, smooth upland sumac, and dwarf sumac; squaw berry, squaw bush, and lemonade sumac. (The name sumac and its various spellings and pronunciations—sumach, shumac, shumack, summaque, and shoemake—are said to be of Arabic origin.)

INDIAN NAMES

The Kiowa call smooth sumac "maw-kho-la" (tobacco mixture) (Vestal and Schultes, 1939, p. 37). The Dakota name is "chan-zi" (yellow-wood), the Winnebago name is "haz-ni-hu" (water-fruit bush), the Pawnee name is "nup-pikt" (sour top), and the Omaha and Ponca name is "chan-zi" (yellow-wood) (Gilmore, 1977, pp. 47, 48). The Lakota call fragrant sumac (*Rhus aromatica*) "can-un'kcemna," which refers to a bush with a bad smell (like human waste) (Rogers, 1980a, p. 43). The Kiowa call it "dtie-aipa-yee-'go" (bitter red berry) (Vestal and Schultes, 1939, p. 39).

SCIENTIFIC NAME

Rhus glab'ra L. is a member of the Anacardiaceae (Cashew Family). *Rhus* comes from the Greek "rhous," which is the name for a bushy sumac. The species name *glabra* means "smooth," in reference to the stems and leaves of the plant.

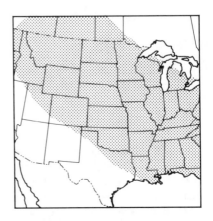

DESCRIPTION

Shrubs 3–5 m (9–15 ft) tall, forming dense thickets. Leaves alternate, pinnately compound, with 11–31 leaflets, each lance-shaped to somewhat oval, 7–9 cm (2¾–3½ in) long, upper surfaces dark green and shiny, margins toothed. Flowers small, in large, branched groups at ends of branches, sometimes male and female separate, in May and Jun; petals 5, greenish. Fruits fleshy, rounded, 3.5–4.5 mm (⅛–³⁄₁₆ in) in diam, red, and hairy, ripening in Aug and Sep.

R. *aromatica* differs in having leaves with only 3 leaflets and yellow flowers. The fruits are very similar to those of R. *glabra*.

HABITAT

Upland prairies, pastures, borders and openings of woods, country roads, and along railroads.

Red ripe berries (summer to fall)—
raw, dried, mostly used for tea or
sumac-ade, or for seasoning; roots
and shoots (spring)—peeled and
eaten raw.

FOOD USE

Smooth and aromatic sumac have
a wide range of similar uses. Several parts have been used medicinally; the roots were used for a
dye; the stems for basketry; the
leaves as a source of tannin for
tanning leather and dried leaves
for smoking mixtures; the berries
for tea; the entire plant as an attractive ornamental; and the
roots, shoots, and berries for food.
Sumac was used by a number of
the Indian tribes of the prairies.
Dr. V. Havard, in "Drink Plants of
the North American Indians"
(1896, p. 44), reported that the tart
fruits of smooth sumac were
bruised in water to make them
"more cooling, refreshing and palatable."

Comanche children were very
fond of the sour, acid fruits of
smooth sumac (Carlson & Jones,
1939, p. 527). There are a number
of wild edible prairie plants (such
as sumac and wood sorrel) that in
our culture and Indian culture are
considered to be "children's
foods." Children are encouraged to
play with and eat these plants because they are not unpalatable or
poisonous, have an unusual or distinct taste, and are not needed as
major food sources.

The Iroquois, who live in New
York, ate smooth sumac sprouts
raw in the spring after they had
emerged (Fenton, 1968, p. 92). To
my sense of taste, the stems of
peeled young sprouts are sweet
and similar to cooked milkweed,
perhaps because of their similar
milky sap. It has also been reported that the peeled roots can be
eaten raw (Morton, 1963, p. 326).
Smooth sumac contains significant quantities of tannin, but apparently not in the young shoots
or peeled roots.

The fragrant sumac, *R. aromatica*, is considerably smaller and
more drought tolerant than
smooth sumac. Its red berries
were used similarly as a food
source. The Kiowa ate berries of
fragrant sumac, which was recognized as one of their "ancient"
foods (Vestal and Schultes, 1939,
pp. 40, 72). They were reported to
eat the berries mixed with corn
meal, beaten with sugar (after exposure to the ways of white men),
or boiled into a tea. It is also possible that this plant may be that
for which one of the six Kiowa societies is named "Ta'-aipeko"
(berries) (Vestal and Schultes,
1939, p. 39).

The antiquity of sumac as a
food is substantiated by analysis
of a preserved human coprolite
from the Ozarks (adjoining the
Prairie Bioregion on the southeast). When the remains of an
Ozark bluff dwelling were excavated, the body of an elderly
woman was found buried. The

fruit of sumac was found to be the principal food that she had last consumed. These remains could not be dated, but it was suggested that they were from at least 20 or 30 centuries ago (Wakefield and Dellinger, 1936, pp. 1412, 1413).

An interesting use of sumac was as a chew-stick to clean teeth. The smooth sumac was used medicinally for treating sore mouths and tongues by Indians, and in the rural Ozarks, it has been used as a chew-stick (Elvin-Lewis, 1979, p. 445). Smooth sumac collected from the Missouri Ozarks was found to contain a highly active antibiotic substance, which is effective in preventing tooth decay (Lewis and Elvin-Lewis, 1977, pp. 218, 238). Chew-sticks can be made easily by cutting off a small stem several inches long, removing the outer bark, and chewing on the tip to soften the fibers, which then can be effectively used to massage the gums. Dogwood species Cornus florida L. and C. drummondii C. A. Meyer can be used similarly, but they have softer fibers and do not taste quite so bitter.

Sumac contains many chemical substances, including significant amounts of tannin. To make tea, perhaps more properly called sumac-ade, without too much tannin, the berries should be removed from their stems and soaked in cold water overnight. Hot water can be poured over the berries, but this will release more of the bitter-tasting tannin. Sumac berries should not be boiled, because this will destroy most of their flavor and also release more tannin.

One unusual effect of sumac, which may relate to chemical substances within the plant, was reported by Charles Millspaugh, whose *American Medicinal Plants*, first published in 1887, is still considered to be a major work in that field. He said (1974, p. 36):

During the summer of 1879, while botanizing . . . I came into a swarm of furious mosquitoes; quickly cutting a branch from a sumach bush. . . . I used it vigorously to fight off the pests . . . (it was) in constant motion (and I was) perspiring freely during the time . . . also ate of the refreshing berries. On three successive nights following this occurence I flew (!) over the city of New York with a graceful and delicious motion I would give several years of my life to experience in reality. Query: Did I absorb from my perspiring hands sufficient juice of the bark to produce the effect of the drug, or was it from the berries I held in my mouth. I noticed no other symptoms, and never before or since enjoyed a like dream.

Smooth and aromatic sumacs are woody shrubs that invade prairies and mismanaged rangeland by spreading from their roots. Lewis and Clark reported first seeing aromatic sumac when journeying up the Missouri River from St. Louis, around the mouth of the Kansas River. On October 1, 1804, they reported from where the Cheyenne River meets the Missouri (in central South Dakota) that sumac is very common in colonies "which appear on the steep declivities of the Hills where they are sheltered from the ravages of the fire" (Thwaites, 1905, 6: 156). Fire, browsing by animals, and the presence of a healthy, tough prairie sod were the primary reasons for sumac not being even more extensive in the original, native prairies.

CULTIVATION

Smooth sumac is an attractive ornamental plant. It should be cultivated for its beauty and for its wide array of uses. Besides the showy red berries in the fall, the leaves turn from dark green to a brilliant red. One cultivar of smooth sumac, called "Flavescens," has yellow foliage in the fall. Aromatic sumac is a smaller shrub whose fragrance some find attractive and others offensive or "skunky smelling." It also has attractive fall foliage. Both varieties can be grown from seeds or transplanted root cuttings.

Ribes odoratum
Buffalo Currant

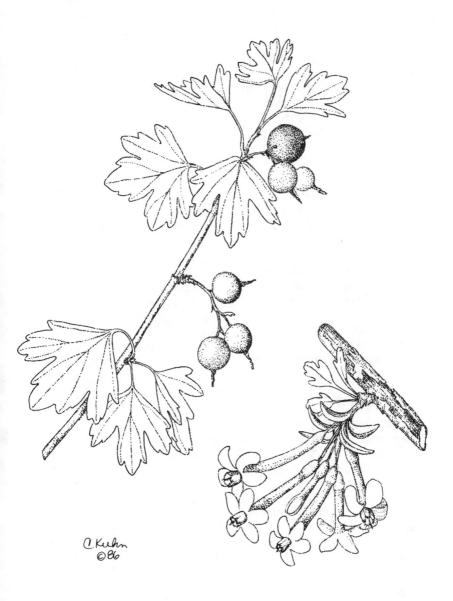

C. Kuhn
©86

Buffalo currant, clove currant, black currant, Missouri currant, golden currant, and flowering currant. (The name clove currant is in reference to the aromatic, clove-like smell of the flowers.)

INDIAN NAMES

The Cheyenne call the black currant "soh'kotasi-mins" (slender, heart-shaped berry) (Grinnell, 1962, p. 175). The Kiowa call it "awdl-kno-bawg" (goose-berry) (Vestal and Schultes, 1939, p. 29). The Dakota, Omaha, and Ponca had names for the similar black currant, *Ribes americanum* Mill. The Dakota call it "wazhushte-cha" (beaver berries) and the Omaha and Ponca name is "pezi nuga" (male gooseberry) (Gilmore, 1977, p. 32).

SCIENTIFIC NAME

Ri'bes odora'tum Wendl. is a member of the Saxifragaceae (Saxifrage Family). *Ribes* may come from the Danish "ribs," referring to the red currant. The species name, *odoratum*, comes from the Latin word meaning "fragrant," in reference to the flowers.

DESCRIPTION

Shrubs 1–2 m (3–6 ft) tall, branches erect to arching. Leaves alternate or in clusters, roundish, 2–5 cm (¾–2 in) long, 3-lobed, margins sometimes coarsely

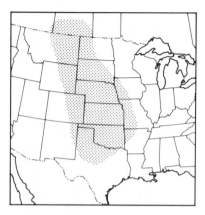

toothed. Flowers fragrant, in elongated groups of 3–8 among leaves, in Apr or May; sepals yellow, alternating with 5 petals, 2–3.5 mm (¹⁄₆—⅛ in) long, yellow to bright red, margins irregular. Fruits fleshy, roundish, 7–9 mm (⁵⁄₁₆–⅜ in) in diam, yellowish, then turning black when ripe in Jul and Aug; seeds egg-shaped, each with a longitudinal rib.

HABITAT

Hillsides, limestone cliffs, and borders of woods, often in sandy areas.

PARTS USED

Ripe berries (summer)—raw, cooked, dried, or made into juice; leaves (summer)—cooked (with meat) or dried for tea.

FOOD USE

The buffalo currant is a sweet and flavorful fruit of the prairies. It was used by many of the Indian tribes and early settlers, and is still used today.

The Kiowa, during the late 1930s, ate the fruits raw and also made them into jelly. They also used this plant as a snakebite remedy because they believed that snakes were afraid of and kept away from it (Vestal and Schultes, 1939, p. 29).

Buffalobird Woman, a Hidatsa, stated that the black currant, *R. americanum*, which is similar to the buffalo currant, was a desirable wild fruit. It was collected in midsummer at the same time as serviceberries, *Amelanchier alnifolia*, and eaten fresh. Currants were dried only when a few were mixed in with serviceberries while collecting. She also reported that young men used the juice of currants to color clay for personal adornment (Nickel, 1974, p. 72). Chokecherries, *Prunus virginiana*, and serviceberries were the main fruits used in making the Indians' dried meat staple called pemmican, but currants were probably also used for this purpose.

Early explorers of the region also appreciated the fruits of the currant bushes. While in Montana on his exploration of the Missouri River region, Prince Maximilian of Wied wrote: "There was an undergrowth of blackcurrants, in search of which our people always went, whenever they had a moment to spare" (Thwaites, 1906, 26: 32). The artist George Catlin also reported during the 1830s (while traveling between the mouth of the Yellowstone River and the Mandan villages in central North Dakota) that there are "wild currants, loaded down with fruit" (1973, 1: 72).

Josiah Gregg, who promoted the Santa Fe Trail, stated in his 1844 *Commerce of the Prairies* that "upon the branches of the Canadian, North Fork, and Cimarron, there are, in places, considerable quantities of excellent plums, grapes, choke-cherries, gooseberries, and currants—of the latter there are three kinds, black, red, and white" (Moorhead, 1954).

To my sense of taste, our native buffalo currant is sweet and flavorful when ripe, although Henry H. Rusby of the Columbia College of Pharmacy in New York apparently did not share my enthusiasm. Writing a series of articles on wild foods in the 1906 *Country Life in America*, he reported that "the delightfully fragrant, yellow-flowering currant of our gardens is native eastward from Colorado, where it produces a nearly worthless black fruit."

The tasty berries of the buffalo currant were used by early settlers because they were one of the few wild fruits found on the prairies. Many people still pick the ripe fruits to make them into jelly or wine. The leaves of some species have been reported to be cooked with meat and also dried and used for tea (Morton, 1963, p. 326).

Currants are an excellent source of vitamin C; a European species was relied on heavily in Britain during World War II when other sources of vitamin C were hard to obtain. The black European currant has four times more vitamin C than an equal amount of oranges (Watt and Merrill, 1963, p. 29). It is probable that the buffalo currant and the other American species of currants also contain significant quantities of vitamin C.

CULTIVATION

The buffalo currant is desirable for cultivation because of its fruit and its showy, golden yellow flowers. While there have been no reports of Indians cultivating or propagating currants, it is possible that they did spread their seeds, either intentionally or unintentionally, to favorable habitats. Near the Pueblo ruin, El Cuartelejo, adjacent to present-day Scott County Lake in western Kansas, there are several large patches of buffalo currants.

Black currants were one of the first fruits grown by my great-great-grandparents on their farm, which was homesteaded in 1871 near Guide Rock, Nebraska. Today, the most common cultivar of buffalo currant is called "Crandall." It has fruit up to ¾ inch in diameter and was found growing wild by R. W. Crandall of Newton, Kansas (Stevens, 1961, p. 302). When dormant, currants can be successfully transplanted to a sunny location. (It should be noted that currants are an alternate host of the white pine blister rust, and in some states their production and shipment are regulated or prohibited.)

Rosa arkansana
Wild Rose

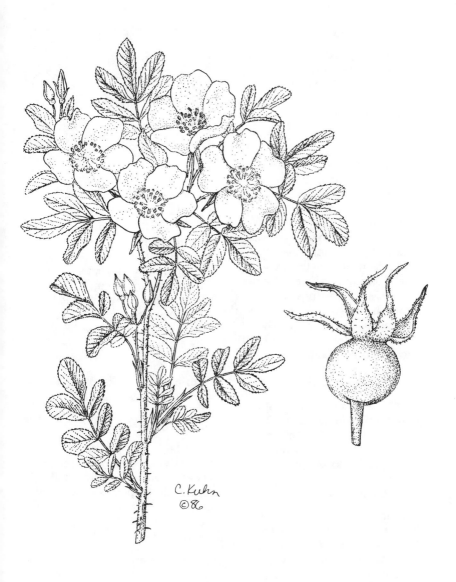

C. Kuhn
©86

Wild rose, prairie wild rose, sun-
shine rose, Arkansas rose,
meadow rose, and pasture rose.

INDIAN NAMES

The Dakota name is "onzhin-
zhintka" (rosebush). The Omaha
and Ponca name is "wazhide" (no
translation given) and the Pawnee
name is "pahatu" (red) (Gilmore,
1977, p. 33). The Cheyenne name
for the rose hip is "hih'-nin"
which means "to pour out" (as
water, flour, grain). This is also
their name for the tomato (Grinnell, 1962, p. 177). The Blackfoot
names are "kini" (rose berries) or
"apis-is-kitsa-wa" (tomato flower)
(Johnston, 1970, p. 314). The
meaning of these comparisons to
tomato is not clear.

SCIENTIFIC NAME

Ro'sa arkansan'a Porter is a
member of the Rosaceae (Rose
Family). *Rosa* is the Latin name
for rose. The species name *arkan-
sana* means "of Arkansas."

DESCRIPTION

Shrubs 1–5 dm (4–20 in) tall,
sometimes dying back to ground
each year, branches with slender
prickles. Leaves alternate, pinnately compound, leaflets 9–11,
egg-shaped to elliptic, 1–4 cm (³⁄₈–
1⅝ in) long, lower surfaces usually hairy, upper ⅔ of margins
toothed. Flowers in groups of 3 or
more at ends of branches from

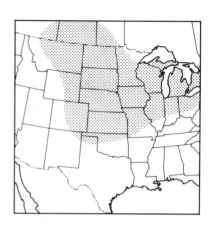

May to Aug; petals 5, separate,
rounded, 1.5–2.5 cm (⅝–1 in)
long, pink to white, rarely deep
rose. Fruits dry, plump, with hairs
along 1 side, 15–30 enclosed in
fleshy covering (hip) that turns red
when ripe in late Aug to Sep.

HABITAT

Prairies, open banks, bluffs, thickets, along roads and railroads.

PARTS USED

Fruit (fall)—raw, stewed, or made
into jelly; young shoots (spring)—
cooked; young leaves and stalks—
for tea; flower petals—raw (in salads).

FOOD USE

The wild rose of our prairies is a
beautiful plant. Its fruits (sometimes called rose hips or berries)
make a good tea that is high in vitamin C. These fruits were also an
emergency food source of the Prairie Indians. When rose hips are

eaten or used for tea, they should be strained to remove the hairs and seeds, which can be irritating to the stomach.

The Hidatsa have a traditional story, called "Picking June Berries," that refers to their use of rose hips. In the story it becomes clear how the Hidatsa view rose hips and their western neighbors, the Crow: "We Hidatsa raise corn, beans, sunflower seed, and good squashes to eat. We are not starving, that we must eat rose berries. . . . Now the Crow Indians like to eat rose berries, and gather them to dry for winter as we dry squashes. We Hidatsa eat rose berries sometimes, but we never dry them for winter. We think they are food for wild men" (Wilson, 1981, pp. 106, 107).

The Assiniboin gathered rose hips in the summer after the berry season, while there was a lull in activities before buffalo hunting time in the early fall. "During this time, the old women picked ripe rosehips. These were washed, mixed with tallow, and allowed to harden. The mixture was eaten with the regular meal" (Kennedy, 1961, p. 83). When there was a shortage of food, the Assiniboin used rose hips as an emergency ration: "When no meat can be found they eat up their reserve of dried berries, pomme blanche and other roots, then boil the scrapings of rawhide with the buds of the wild rose, collect old bones on the prairie, pound them and extract the grease by boiling. A still

greater want produces the necessity of killing their dogs and horses for food, but this is the last resort and approach of actual famine, for by this they are destroying their means of traveling and hunting" (Denig, 1928, p. 509).

The Blackfoot ate the fruits fresh or roasted, after the seeds were removed (Johnston, 1970, p. 314). Sometimes these rose hips were crushed and added to their pemmican. The fruits were also used to make necklaces before trade beads were acquired (ibid.). In the winter, rose hips could still be found clinging to rose bushes and were used as a famine food (Hellson, 1974, p. 105).

The Cheyenne infrequently ate the fruit and the petals. They state that rose hips should not be eaten too freely and some suggested that they caused "itchy buttocks" (Hart, 1981, p. 36).

Rose hips were also an emergency food of the Pawnee, Omaha, Dakota, and the Ponca (Gilmore, 1977, p. 33). In 1821 during the Bucks-Rattling-Antlers-Moon (in early fall), the Osage were struggling against starvation and wild rose fruits were one of the foods that the women and children gathered (Matthews, 1961, p. 489). Wild roses can be counted on as an emergency food during droughts because they are one of the deepest-rooted prairie plants. John Weaver (1974, p. 26) found that the taproot of the wild rose reached to a depth of over 21 feet.

Melvin Gilmore states in *Uses*

of *Plants by the Indians of the Missouri River Region* (1977, p. 33) that the Dakota and other tribes think of animals and plants as having their own songs. They believe that music and song are "an expression of the soul and not a mere artistic exercise." This belief is based on the idea of earth (the Mother Earth) as having a living consciousness—a concept that has recently gained greater acceptance in our own culture with the teaching of ecological concepts. The Dakota "truly venerated and loved" the earth; they "considered themselves not as owners or potential owners of any part of the land, but as being owned by the land which gave them birth and which supplied their physical needs from her bounty and satisfied their love of the beautiful by the beauty of her face in the landscape" (Gilmore, 1977, p. 33).

The following is a translation by Dr. A. McG. Beede of an old Dakota song (ibid.). The word "Mother" in the song refers to "Mother Earth." The first stanza is an introduction by the narrator. The trilled musical syllables at the close of the last two stanzas express the spontaneous joy of a person who has "life-appreciation of Holy Earth."

Song of the Wild Rose

*I will tell you of something I
  know,
And you can't half imagine how
  good;*

*It's the song of wild roses that
  grow*

*In the land the Dakota-folk love.
From the heart of the Mother we
  come,
The kind Mother of Life and All;
And if ever you think she is
  dumb,
You should know that flowers are
  her songs.*

*And all creatures that live are her
  songs,
And all creatures that die are her
  songs,
And the winds blowing by are her
  songs,
And she wants you to sing all her
  songs.*

*Like the purple in Daydawn we
  come,
And our hearts are so brimful of
  joy
That whene'er we're not singing
  we hum
Ti-li-li-li-i, ta-la-la-loo, ta-la-la-
  loo!*

*When a maiden is ready to wed
Pin wild roses all over her dress,
And a rose in the hair of her
  head;
Put new moccasins onto her feet.
Then the heart of the Mother will
  give
Her the songs of her own heart to
  sing;
And she'll sing all the moons she
  may live,
Ti-li-li-li-i, ta-la-la-loo, ta-la-la-
  loo!*

Wild roses were often mentioned by the early explorers and travelers in the Prairie Bioregion. During the 1830s, while George Catlin was between the mouth of the Yellowstone River and the Mandan villages, he reported that the "wild rose bushes were decked out in all their glory" (1973, p. 72). In 1820, Dr. E. James of the Long Expedition reported seeing rose bushes near a Pawnee Indian village along the Loup River (in eastern Nebraska): "The shore, opposite the Loup village, is covered with shrubs and other plants, growing among the loose sands. One of the most common is a large flowering rose, rising to about three feet high, and diffusing a most grateful fragrance" (McKelvey, 1955, p. 215).

There are many edible parts of the wild rose, besides the fruits. The young shoots in spring are said to make an acceptable pot herb, and the roots (in addition to the stems) can be used for making tea (Harrington, 1967, p. 272). Rose bark was a favorite of the Hidatsa for tea. They removed the red outer bark and boiled the inner bark. It tastes much like commercial black tea (Weitzner, 1979, p. 214). Also, the petals are an attractive garnish for a salad, can be used to make tea, or utilized strictly for their pleasant flavor and fragrance added to butter, hair oil, or perfume.

Rose hips vary widely in flavor. They need to be picked just when they are ripe, although a light frost may sweeten them. Dr. V. Havard of the U.S. Army reported (1895, p. 122) that "the fruit or hip of several of our wild Roses, after being touched by frost, is sweet and palatable." But regardless of the time of harvest or their color, some rose fruits are surprisingly bland, while others are tasty, sweet, and acidic.

Rose hips are one of the best natural sources of vitamin C (ascorbic acid) and are used in making vitamin pills. The vitamin C content of rose hips varies widely. They have been reported to have up to 1750 milligrams of vitamin C per 100 grams (Arnason, 1981, p. 2239), whereas an equal amount of oranges has only 71 milligrams (Watt and Merrill, 1963, p. 41). Three rose hips have as much vitamin C as a whole orange (Phillips, 1979, p. 123).

I have included this comparison to point out that the basic requirement for vitamin C does not have to be fulfilled by importing citrus fruits. In *The Kansas Food System—Analysis and Action Toward Sustainability* (1982, p. 40) I estimated that for the year 1980, over $23 million worth of citrus fruit was imported into Kansas. It would be interesting to see how popular rose hips and other local foods would become if their nutritional values were promoted as well as those of citrus fruits.

CULTIVATION

Wild roses were probably not cultivated by native peoples of the Prairie Bioregion. Dr. E. James's observation of the wild roses opposite the Loup River from a Pawnee village may indicate encouragement of this species in suitable sandy habitats. The high traffic and use of the land around village sites altered the native vegetation. It would seem that any plants observed near the villages, as wild roses were, should be considered to be aggressive, tolerant, or encouraged.

Wild roses are somewhat difficult to propagate. Seeds should be sown or statified soon after harvest, but many will not germinate until the second year. Transplanting is often not successful—it is hard to move a root that is 21 feet deep! Divisions or cuttings of the roots may occasionally be successful.

Rubus flagellaris
Dewberry

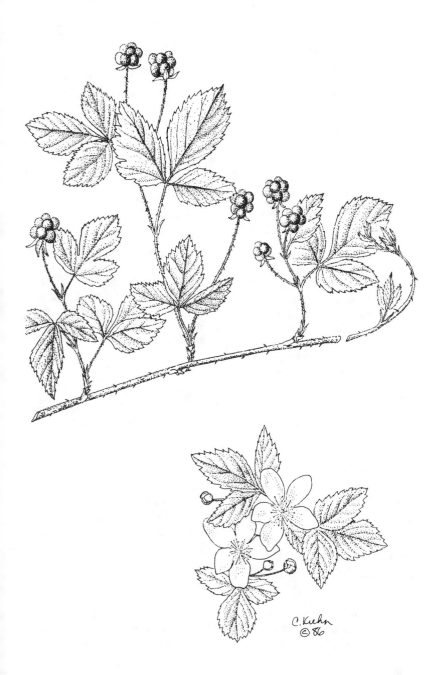

COMMON NAMES

Dewberry, highbush blackberry, and blackberry.

INDIAN NAMES

The Plains Cree call the dewberry "osksikomina" (eye berry or dew berry) (Mandelbaum, 1940, p. 203).

SCIENTIFIC NAME

Ru'bus flagellar'is Willd. and *R. ostryifol'ius* Rydb. are members of the Rosaceae (Rose Family). *Rubus* is derived from the Latin word "ruber," which means red, in reference to the color of the fruits and stems of some species. The species name *flagellaris* means "whip," suggested by the terminal whiplike stems, and *ostryifolius* means "leaves like *Ostrya*," a member of the Birch Family.

DESCRIPTION

Woody perennials with trailing branches 0.5–5 m (1½–15 ft) long, often rooting at tips, armed with curved prickles. Leaves alternate, compound with 3 or 5 leaflets, egg-shaped to elliptic, 2–7 cm (¾–2¾ in) long, surfaces sometimes hairy, margins toothed. Flowers in flat-topped or elongated groups or sometimes solitary on erect branches, from Apr to Jun; petals 5, separate, egg-shaped, 1.5–3 cm (⅝–1¼ in) long, white or rarely pinkish. Fruits fleshy clusters of several segments, each containing 1 seed, roundish, 1–2.5 cm (⅜–1

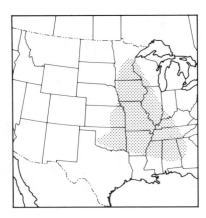

in) long, black, juicy, and sweet, separating from fleshy central columns when they fall.

R. ostryifolius differs in being taller (to 2.5 m or 8 ft) with more erect branches and in having fruits that remain attached to the fleshy central columns when they fall.

HABITAT

Rocky open woods, thickets, prairies, along roadsides, and railroad embankments; also in pastures and fields and along fencerows.

PARTS USED

Fruit (in summer)—raw, cooked, or preserved; leaves (early spring)—for tea; young, tender shoots (early spring)—peeled and eaten raw.

FOOD USE

The dewberry and blackberry are well know for their tasty fruits. They are native to the southern and eastern portion of the Prairie

Bioregion and are part of the woody/brushy transition from prairie to woods. Berry patches and thickets gradually replace the prairie grasses and forbs, unless fire or other disturbances reverse the process. Then these berry patches and thickets are gradually replaced by woody shrubs and trees.

The dewberry can be distinguished from the blackberry by its trailing or sprawling stems. The Plains Cree gathered the edible berries of both species (Mandelbaum, 1940, p. 203) and they were undoubtedly eaten when ripe by all the Plains Indian tribes, wherever they could be found. Seeds of *Rubus* species have been recovered from the Seven Acres and Maybrook sites of villages around the twelfth and thirteenth centuries A.D. in Jackson County, Missouri, near Kansas City (Adair, 1984, pp. 35, 70).

The abundance of berries and fruits was mentioned by some of the early explorers and travelers of the region. Edmund Flagg reported while traveling in southwestern Illinois in 1836 or 1837, near the confluence of the Kaskaskia and Mississippi Rivers: "The ride from Kaskaskia to Prairie du Rocher in early autumn is truly delightful . . . the path lay through a tract of astonishing fertility, where the wild fruit flourishes with a luxuriance known to no other soil. Endless thickets of the wild plum and the blackberry, interlaced and matted together by the young grape-vines streaming with gorgeous clusters, were to be seen stretching for miles along the plain. Such boundless profusion of wild fruit I had never seen before" (Thwaites, 1906, 27: 69).

Edwin Palmer reported on the use of dewberry and blackberry in his "Food Products of the North American Indians" (1871, p. 415):

Common blackberry—Found in Northern Missouri, Texas, California, and Minnesota. The Indians keep in remembrance the localities where this plant grows, and are as fond of its fruit as are the whites.

Dewberry—Grows abundantly in Southern Kansas, having a fine rich flavor, and is held as a great delicacy by Indians and whites.

The tasty fruit of the blackberry is not its only gift; its dried leaves make a fine tea. The leaves should be picked in the early spring and dried for later use (Phillips, 1979, p. 27). Also, the young shoots can be eaten raw after being gathered early in the spring and peeled (Morton, 1963, p. 326). I have found young, peeled blackberry shoots to be sweet but pithy and a little tough.

CULTIVATION

Numerous varieties of dewberries and blackberries can be cultivated. Horticultural varieties of dewberries have been developed from wild plants. One variety, Bartles' Mammoth Dewberry, "which appeared spontaneously in an abandoned cornfield in southern Illinois, had large, rich, and juicy berries, the offspring of the original plants yielding 60–80 bushels per acre" (Stevens, 1961, p. 250). Horticultural varieties of black-berries are even more productive than dewberries. Dewberries and blackberries can be propagated by root cuttings, by transplanting suckers, or the tip of the cane or a node can be covered with soil and roots will develop.

Shepherdia argentea
Buffalo Berry

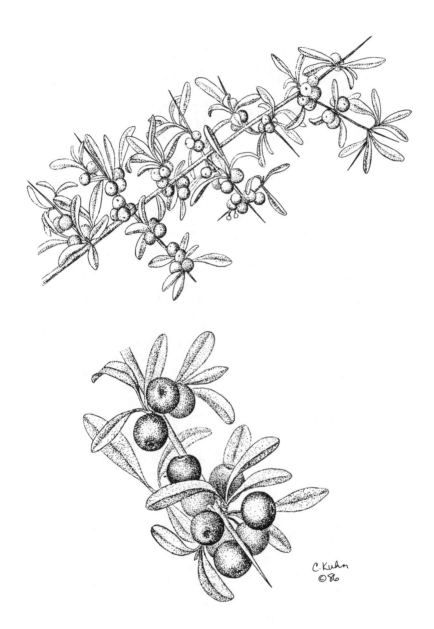

C. Kuhn
©86

Buffalo berry, silver buffaloberry, thorny buffaloberry, bullberries, and "graise de boeuf." (This last name is the French voyageurs' name. All of these names were given because the buffalo were fond of this shrub [Johnston, 1970, p. 316]).

INDIAN NAMES

The Blackfoot name is "miksinit-sim" (bull berry) (Johnston, 1970, p. 316). The Lakota name is "mas'tinca-pute'-can" (rabbit lip tree) (Rogers, 1980a, p. 68). The Cheyenne name is "mat'si-ta-si-'mins" (red hearted) (Grinnell, 1962, p. 181). The Winnebago name is "haz-shutz" (red-fruit); the Omaha and Ponca name is "zhon-hoje-wazhide" (gray something); and the Pawnee name is "karitsits" (Gilmore, 1977, p. 54). The Assiniboin name is "tasque-sha-shah" (Denig, 1930, p. 583). The Arikara name is "nar-nis" and the Mandan name is "as-say" (Thwaites, 1904, 1: 160n). The Crow name is "ingahawmp" (Blankenship, 1905, p. 24).

SCIENTIFIC NAME

Shepherd'ia argen'tea (Pursh) Nutt. is a member of the Elaeagnaceae (Oleaster Family). *Shepherdia* is named for the English botanist John Shepherd. The species name *argentea* means "silvery," referring to the leaves.

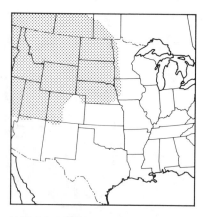

DESCRIPTION

Erect shrubs of small trees, 1–6 m (3–18 ft) tall, some twigs ending in spines, young branches hairy, older branches dark gray. Leaves opposite, oblong and rounded at tips, 2–5 cm (⅜–2 in) long, gray green, surfaces covered with silvery scales. Flowers in small clusters on younger branches, in May and Jun, male and female separate; sepals 4, brown, erect or spreading, 1–3 mm (⅙–⅛ in) long, no petals. Fruits fleshy, egg-shaped, 5–7 mm (³⁄₁₆–⁵⁄₁₆ in) long, red, ripening in Jul and Sep.

HABITAT

Prairie valleys; banks of streams; steep, eroded, dry hillsides (often in poor soil).

PARTS USED

Ripe fruit (fall, sweetened by frost)—raw, cooked, or dried, also used to make juice.

The buffalo berry is found in the northern half of the Prairie Bioregion. The fresh fruits are tart, but pleasant. They have a fairly large seed, but it is easily chewed up.

The Blackfoot gathered the red buffalo berries after the first frost and they were eaten fresh or dried for winter use (Johnston, 1970, p. 316). When the berries were ripe and juicy, the Blackfoot "sometimes mashed them with a stick in a buffalo horn and drank the juice from the horn" (Ewers, 1958, p. 87). Buffalo berry juice is reported to be a delicious summer drink (Fernald et al., 1958, p. 277).

The Assiniboin also gathered buffalo berries, which they first washed and then dried. Sometimes the berries were added to a soup made from meat broth (Kennedy, 1961, p. 83). Buffalobird Woman reported that the Hidatsa collected buffalo berries in the fall after a hard frost and that they consumed them fresh, but not dried or preserved (Nickel, 1974, p. 73).

George Catlin, who painted many activities in the life of the Mandan, was served a "paste or pudding" made by them of prairie turnip (*Psoralea esculenta*) flour and "finely flavored with buffalo berries which are collected in great quantities in this country" (1973, 1: 115). The Reverend J. Owen Dorsey reported (1881, p. 306) that the Omaha ate buffalo berries raw, or they dried and boiled them before eating. The Dakota (Sioux) ate the fruits fresh or dried and ceremonially used the fruits of chokecherry (*Prunus virginiana*) or buffalo berry at feasts given in honor of a girl reaching puberty (Gilmore, 1977, p. 54). It is possible that these red fruits were used because their color symbolized menstruation.

George Bird Grinnell reported (1962, p. 181) on the use of buffalo berries by the Cheyenne and other Indian tribes: "The buffalo-berry—often called the bull-berry—does not grow very freely in the present country of the Northern Cheyenne; but so far as they collect it, this is done much as the Indians of the Missouri River gather bull-berries. When the berries are ripe, robes or skins are placed on the ground, and the thorny bushes are beaten with sticks so that the berries fall from the twigs and may be gathered up on the skins." Buffalo berries were also reported to have been used by the Pawnee, Ponca, Winnebago (Gilmore, 1977, p. 54), and Crow (Blankenship, 1905, p. 24).

The buffalo berry, which grew in such profusion downstream from the mouth of the Yellowstone River on the Missouri River, was described in 1833 by George Catlin in his *Letters and Notes on the Manners, Customs, and Conditions of the North American Indians* (1973, 1:74).

The buffalo berries, which are peculiar to these northern regions, lined the banks of the river and defiles in the bluffs, sometimes for miles together; forming almost impassible hedges, so loaded with the weight of their fruit, that their boughs were everywhere gracefully bending down and resting on the ground.

This shrub, which may be said to be the most beautiful ornament that decks out the wild prairies, forms a striking contrast to the rest of the foliage, from the blue appearance of its leaves, by which it can be distinguished for miles in the distance. The fruit which it produces in such incredible profusion, hanging in clusters to every limb and to every twig, is about the size of ordinary currants, and not unlike them in colour and even in flavor; being exceedingly acid, and almost unpalatable, until they are bitten by the frost of autumn, when they are sweetened, and their flavour delicious; having, to the taste, much the character of grapes, and I am inclined to think, would produce excellent wine.

Catlin's traveling partners, Bogard and Batiste, became excited when they contemplated making wine from buffalo berries. The three of them agreed that two men could gather 30 bushels of the fruit per day. He continued (ibid., pp. 74–75):

We several times took a large mackinaw blanket which I had in the canoe, and spreading it on the ground under the bushes, where they were the most abundantly loaded with fruit; and by striking the stalk of the tree with a club, we received the whole contents of its branches in an instant on the blanket, which was taken up by the corners, and not unfrequently would produce us, from one blow, the eighth part of a bushel of this fruit; when the boughs relieved of their burden, instantly flew up to their native position.

Dr. Ferdinand V. Hayden made geological explorations of the northern prairies in the 1850s and 1860s. He reported (1859, p. 742) that the buffalo berry was very abundant along the Missouri River from the mouth of the Sioux River (which divides South Dakota and Iowa) all the way to the mountains (in Montana). He further stated that the buffalo berry, when dried, takes the place of chokecherries in pemmican and that along with chokecherries and wild plums, it is a principal food of the bears and wolves (Hayden, 1862, p. 370).

CULTIVATION

The buffalo berry with its silvery foliage and red berries is an attractive plant, and because its fruit is so tasty, it should be considered for cultivation. J. W. Blankenship, in "Native Economic Plants of Montana" (1905, p. 24) reported that the buffalo berry was an important food of the Indians and the French voyageurs. "Nor has its value diminished with the coming of civilization for they are still found in our autumn markets and are highly esteemed for jelly, the demand exceeding the supply. If the larger, sweeter, yellow colored variety could be made thornless, it would become one of the most valued fruits of the orchard for the whole semi-arid region of the plains."

The buffalo berry is adapted to dry and rocky soils and cold temperatures. It is used as a hedge in the northwestern United States, where it is valued for its great hardiness. A cultivar with yellow fruit is named "Xanthocarpa." Buffalo berries can be propagated by planting nursery stock, by transplanting shrubs from the wild, or from seeds sown outdoors in the fall or stratified and planted in the spring.

Stanleya pinnata
Prince's Plume

C. Kuhn
©86

COMMON NAMES

Prince's plume, desert plume, Indian cabbage, and yellow poker.

INDIAN NAMES

None were found in the sources consulted.

SCIENTIFIC NAME

Stan'leya pinna'ta (Pursh) Britt. is a member of the Brassicaceae (Mustard Family). *Stanleya* is in honor of Lord Edward Stanley, an Englishman who was a president of the Linnean Society. The species name *pinnata* means "feathered," in reference to the divisions sometimes present on the lower leaves.

DESCRIPTION

Shrubby perennials to 1.5 m (5 ft) tall, with several stems from a woody base, branched above. Leaves alternate, oval to somewhat linear, 13–20 cm (5–8 in) long, thick, usually entire but lower ones sometimes deeply lobed. Flowers in long, dense, and very showy groups at tops of branches, from Apr to Aug; petals 4, separate, elliptic with narrow bases, 10–15 mm (⅜–⅝ in) long, yellow, bases brownish, 6 stamens much longer than petals. Fruits dry, linear, 2–8 cm (¾–3¼ in) long, curved.

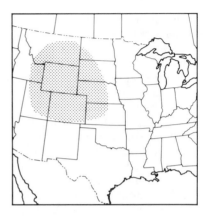

HABITAT

Dry hills and plains; considered to be an indicator of selenium soils.

PARTS USED

Leaves, young plants, young stems (spring)—cooked; seeds (fall)—ground for mush. Not recommended for use because of possibly toxic levels of selenium.

FOOD USE

Prince's plume has been referred to as the "sentinel of the Plains" because of its conspicuous, plumelike spikes of yellow flowers, which especially stand out against a darkened skyline (Rollins, 1939, p. 109). It grows in western prairies and deserts, particularly in areas of poor soil that are often selenium-laden. It can be poisonous because of its ability to accumulate large amounts of selenium in its tissues. It is included here because it was probably a minor food source of Indians of the Prairie Bioregion and is an interesting and unusual plant.

Prince's plume has been called Indian cabbage because it was reported as a cabbagelike food source of the Indians of the Southwest and the Great Basin. The Hopi boiled and ate the greens (Wyman and Harris, 1951, p. 25). Edward Palmer reported (1878, p. 604) that prince's plume is "eaten raw in the spring by the Pah-Ute Indians, the young plants being tender, and when cooked taste like cabbage. For this reason these plants are called cabbage by the settlers of Utah. The Indians gather the seeds and after reducing them to flour make them into mush."

Dr. Henry H. Rusby described (1907, p. 68) prince's plumes as "cabbage-like plants," but said that they were used for food only in emergencies and had "to be specially prepared to remove their disagreeable bitter taste and poisonous properties." Flo Reed reports in *Uses of Native Plants by Nevada Indians* (1970, p. 3) that several waters must be poured off prince's plume; otherwise it acts as an emetic.

Selenium is an essential element for human health; however, it is only needed in small quantities. The poisonous nature of prince's plume comes from the large quantities of selenium that accumulate in its tissues. Al-

though it is not certain whether selenium is necessary for the normal growth of prince's plume, the plant grows in soil with at least small quantities of selenium in one form or another (Rollins, 1939, p. 109). Because of this relationship, prince's plume is used as an indicator plant for selenium soils.

Prince's plume is also poisonous to livestock. The symptoms of this poisoning are listed in *Pasture and Range Plants* (Phillips Petroleum Company, 1959, p. 12): "Symptoms of the first stage are: roughened coat, loss of weight, tendency to stray from main herd, and impairment of vision. The next stage is characterized by pronounced blindness and depraved appetite causing the animal to chew fence rails, bones, wire or any metallic object available. Animals may wander in circles and push forward into any solid objects encountered, such as buildings or fences."

CULTIVATION

Prince's plume is an unusual-looking plant that would be a nice addition to a wildflower garden or prairie restoration. It probably can be propagated from either seed or root division. Because of its affinity to selenium soils, a bucket of the soil in which it is growing should also be transplanted with the plant or mixed into the seed bed.

Tradescantia occidentalis
Spiderwort

C. Kuhn
©86

COMMON NAMES

Spiderwort, prairie spiderwort, spider lily, ink flower, and king's crown. "Spiderwort" is derived from the Anglo-Saxon "wyrt" or "wort," meaning herb or root; "spider" apparently comes from the fact that when the stems are broken and pulled apart, their copious mucilaginous slime is drawn into thin threads, suggesting a spider web (Stevens, 1961, p. 44).

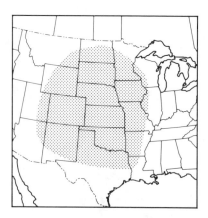

INDIAN NAMES

None were found in the sources consulted.

SCIENTIFIC NAME

Tradescant'ia occidental'is (Britt.) Smyth. is a member of the Commelianaceae (Spiderwort Family). *Tradescantia* was named in honor of John Tradescant, gardener for Charles I of England, who assembled a wealth of plant materials. The name *occidentalis* means "western," referring to the species' distribution west of the Mississippi River (Stevens, 1961, p. 44, 45).

DESCRIPTION

Perennial herbs to 5 dm (20 in) tall, stems erect and often branching. Leaves alternate, linear to lance-shaped, often folded, 9–33 cm (3½–13 in) long, bases wrapped around stems and wider than blades. Flowers lasting 1 day, in flat-topped groups at ends of main stems or branches, from May to Aug; petals 3, broadly egg-shaped, 0.7–1.6 cm (¼–⅝ in) long, pink, purple, or blue. Fruits dry, small, roundish, opening into 3 sections; seeds gray, flat, oblong, and pitted.

HABITAT

Prairies, roadsides, and railroad rights-of-way, particularly in sandy soil.

PARTS USED

Young stems and leaves (spring)— raw in salads or cooked as a pot herb; flowers (summer)—raw to garnish salads.

FOOD USE

Spiderwort with its showy blue flowers is a beautiful plant of the Prairie Bioregion. There are several species, all of which are edible. George Washington Carver, who developed and found many food uses for sweet potatoes and peanuts, also wrote about wild edible plants. He highly recommended spiderwort and described it as "rich flavored" (Fernald et al., 1958, p. 124).

The Cherokee of North Carolina eat the closely related and similar *T. virginiana.* "The young growth is parboiled and then fried, frequently mixed with other greens" (Witthoft, 1977, p. 252). The tender shoots of spiderwort are also good in salads and the flowers can be a colorful, edible garnish. The shoots and tender leaves can also be harvested in the late spring and early summer as a pot herb. They can be steamed or cooked, without changing the cooking water.

Melvin Gilmore, in *Uses of Plants by the Indians of the Missouri River Region* (1977, p. 18) gives a colorful description of spiderwort: "This is a charmingly beautiful and delicate flower, deep blue in color, with a tender-bodied plant of graceful lines. There is no more appealingly beautiful flower on the western prairies than this one when it is sparkling with dewdrops in the light of the first beams of the rising sun. There is about it a suggestion of purity, freshness, and daintiness."

Gilmore also tells of the use of this plant by the Dakota as a love song charm. When a young man of the Dakota Nation was in love, he would walk alone on the prairie and sing songs to the spiderwort, which would personify the girl he loved. "In his mind the beauties of the flower and of the girl are mutually transmuted and flow together into one image" (ibid.).

The following song addressed to the spiderwort was translated from the Dakota by Dr. A. McG. Beede (ibid.):

Wee little dewy flower
So blessed and so shy,
Thou'rt dear to me, and for
My love for thee I'd die.

CULTIVATION

Although spiderwort can be harvested in the wild when large, healthy populations are found, it also can be successfully grown in a garden. Spiderwort can be propagated from seed or a stem cutting, or it can be transplanted to a light or sandy soil. A cultivated variety of *T. occidentalis* called "Rubra" has been developed for its red flowers and there are many cultivars of *T. virginiana.*

Viola pedatifida
Prairie Violet

Prairie violet, violet, purple prairie violet, and larkspur violet.

INDIAN NAMES

None were found in the sources consulted.

SCIENTIFIC NAME

Viol'a pedatif'ida G. Don. is a member of the Violaceae (Violet Family). *Viola* is the classical Latin name for the violet. The species name *pedatifida* means "foot cleft" in reference to the cleft leaves.

DESCRIPTION

Perennial herbs with leaves rising from stout, underground stem bases, no aboveground stems. Leaves alternate, on stalks to 15 cm (6 in) tall, divided into 3 parts, each one cut to base into linear lobes, each of these cut again into 2–4 segments. Flowers solitary on stalks usually taller than leaves, from Apr to Jun; petals 5, 10–18 mm (3/8–3/4 in) long, violet to reddish-violet, lower 3 white at bases and veined with dark purple, upper 2 obviously dark-veined. Fruits dry, oval, 8–12 mm (5/16–1/2 in) long, yellow-green, opening into 3 parts; seeds light brown.

HABITAT

Tallgrass prairie (visible in early spring).

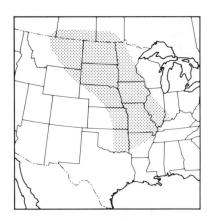

PARTS USED

Leaves (spring)—raw or cooked in soup; flowers (spring)—raw to garnish salads.

FOOD USE

This little-mentioned prairie species is soon dwarfed by other plants in the spring, but not before its attractive flowers are evident. There is no record of prairie violet as a food source, but quite likely its leaves and flowers have been used by the Indians in soups and stews. Also, in the spring when greens were probably in high demand, it may have been eaten in its raw state.

Violets were reported by Melvin Gilmore (1977, p. 51) to be used in a game played by Omaha children: "In springtime a group of children would gather a quantity of violets; then, dividing into two equal parties, one party took the name of their own nation and the other party took another, as for instance Dakota. The two parties sat down facing each other, and each player snapped violets with his opponent till one or the other had none remaining. The party having the greater number of violets remaining, each party having had an equal number at the beginning, was the victor and playfully taunted the other as being poor fighters." Children's games are often very old traditions and seem to have common themes. Julian Steyermark (1981, p. 1071) tells of a surprisingly similar game played in the Ozarks with the pansy violet, *V. pedata.*

It is probable that all species of violets are edible, with some being more palatable than others. Some of the yellow-flowered species have been reported to be cathartic and may purge the bowels (Elias and Dykeman, 1982, p. 95). Fortunately, our prairie species, which have violet flowers, apparently do not cause this reaction. The young, small leaves of violets may be confused with other plants that might not be edible, so it's best to collect these plants when in bloom. It is also reported that the roots of some violets are emetic, so only the leaves and flowers should be used (Fernald et al., 1958, p. 275).

The leaves of wild violets are mucilaginous and were used as a substitite for okra in the South during the Civil War (Fernald et al., 1958, p. 275). The basal leaves of common blue violets, *V. papi-*

lionacea Pursh, were found to be the highest in vitamin A and C of any of the 17 wild edible foods studied by Zennie and Ogzewella (1977, pp. 78, 79) in Ohio and Kentucky. They had more than twice as much vitamin A as an equal weight of spinach and over five times as much vitamin C as an equal weight of oranges.

Violet flowers make colorful garnishes for a spring garden salad. However, our prairie violet is not always common, and its beauty is greater than its food value. Generally, it should be left to grow in its natural habitat.

CULTIVATION

This colorful plant can be established through seed or by division of the rootstalk (take some and leave some) in the late summer (best), fall, or spring. Seeds can be planted outdoors in the fall or in the spring after stratifying. Fresh seed produces the best results. These plants need a moist location and some shade in summer. They are short-lived perennials that may be killed in winter if there is insufficient moisture.

Julian Steyermark suggests that the pansy violet, *V. pedata* L., with its showy flowers—it has two shades of purple petals— should be cultivated in flower gardens: "This violet is a handsome gem for rock-gardens and succeeds in any sunny or semi-shaded, well drained, usually acid soils, where it thrives best on sand, sandstone, chert, granite,and similar acid rocks. It will also grow on limestone. It does best on slopes on soils having good drainage. Along highway cuts through cherty or sandstone substrata, one often sees thousands of these plants growing thickly in rocky or gravelly exposures" (p. 1071).

Prairie violets should also be cultivated. They need a moist location and in their natural habitat, these shallow-rooted plants are usually shaded in the summer. During the drought of the 1930s, John Weaver reported, *Viola* species developed only poorly, and not infrequently the flowers withered and virtually no seeds were produced (1968, p. 147).

Yucca glauca
Small Soapweed

C. Kuhn
©86

Small soapweed, soapweed, soap-well (these three names refer to the fact that the root is used as a soap substitute), yucca, beargrass, New Mexican Spanish pamilla, amole, Spanish bayonet, dagger plant, and Adam's needles (these last three names point out that the leaves are sharp).

INDIAN NAMES

The Lakota name is "hupe'stola" (sharp-pointed stem) (Rogers, 1980a, p. 30). The Blackfoot name is "ek-siso-ke" (sharp vine or snake weed) (Johnston, 1970, p. 309). No translations were given for the following names: Kiowa, "kaw-tzee-a-tzo-tee-a, ol-po-on-a" or "kee-aw-gee-tzot-ha'-a'h" (Vestal and Schultes, 1939, p. 17); and Dakota, "hupestula"; Omaha and Ponca, "duwaduwa-hi"; and Pawnee, "chakida-kahtsu" or "chak-ila-kahtsu" (Gilmore, 1977, p. 19).

SCIENTIFIC NAME

Yuc'ca glau'ca Nutt. is a member of the Lilaceae (Lily Family). *Yucca* was mistakenly given its name by Gerard, who thought a specimen imported from the West Indies was the "yuca" or "manioc," from which tapioca is made. The species name *glauca* means "whitened with a bloom," because the leaves are somewhat whitened by a waxy film that helps to keep them from drying out.

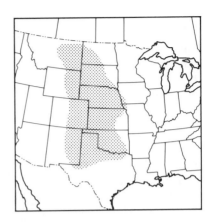

DESCRIPTION

Perennials with thick, under-ground stem-bases, often growing in clumps. Leaves radiating from basal rosettes, linear, 4–7 dm (16–28 in) tall, thick, fleshy, margins whitish and fringed. Flowers in branched groups at tops of stalks extending above leaves, from May to Jul; each one roundish or bell-shaped, with 6 separate segments, thick, oval, 3.5–5.5 cm (1⅜–2¼ in) long, greenish white to cream. Fruits dry, 6-sided, about 4.5–6 cm (1¾–2¼ in) long, opening to release numerous flat, black seeds.

HABITAT

Upland prairies, plains, sandy blowouts, and hillsides (often of limestone soils).

PARTS USED

Flower stalk (spring)—used like asparagus; immature fruits (summer)—cooked; petals (late spring)—raw in salads.

This evergreen perennial is a characteristic species of the sandy, dry areas of the Prairie Bioregion. It was more commonly used as a food plant in the Southwest along with another soapweed, *Y. baccata* Torr., which has a superior flavor. The flower stalk that emerges in the spring can be cooked and eaten like asparagus. The Kiowa ate this central spike, which they referred to as "cabbage" (Vestal and Schultes, 1939, p. 18). The Chiricahua and Mescalero Apache roasted these flower stalks over a bed of embers for about 15 minutes. The inside portion of the stalk was considered the most delicious (Opler, 1936, p. 38).

The cooked flower stalks that I have prepared have had a bitter taste. This could be due to improper preparation, to their flavor not agreeing with my sense of taste, or to variations in taste between plants within their distribution. The latter may be important, because soapweed species, including *Y. glauca*, are quite popular and considered tasty in the Southwest, but are hardly mentioned as a food source in the central and northern portions of their range. It is possible that the climate of the Southwest influences the flavor of this species.

The buds and cream-colored flowers that grow from the stalks are quite tender, but their flavor also seems to vary from plant to plant. I have found some that are quite sweet and others that taste quite soapy and a little bitter. The flowers were eaten by the Mescalero Apache (Castetter, 1935, p. 56).

The immature fruits of soapweed also have been used as food. Matilda Stevenson, in "Ethnobotany of the Zuni" (1915, p. 73), reported their use of the seed pods of soapweed, *Y. glauca:* "The seed pods, which are slightly sweet, are boiled. The young pods are considered far superior as food to the older ones. While the seeds of the former are eaten with the pods, those of the latter are not regarded as edible. These pods are not combined with other foods, and they are never eaten warm or with meals. 'They would not agree with the stomach if taken with other food.'"

Soapweed has many other uses. Bundles of sharp-pointed leaves were bound together and the point was used by the Dakota as a fire drill (to start a fire) (Gilmore, 1913b, p. 358). They used the root, which produces a sudsy soap when soaked in water, for washing the scalp; they also used the root to make a decoction for use in tanning hides (ibid.). Fibers from the leaves were used to weave sandals and other articles and have been found in the centuries-old remains of the Ozark Bluff-dweller culture (Gilmore, 1931b, p. 95). Furthermore, the roots have been used for medicine (Gilmore, 1977, p. 19).

The ecological context of our native soapweed can be understood by Steven Long's 1819 description of it growing in the Sand Hills of west-central Nebraska:

On the summits of some of the dry sandy ridges, we saw a few of the plants called Adam's needles, (Yucca glauca) thriving with an appearance of luxuriance and verdure, in a soil which bids defiance to almost every other species of vegetation. Nature has, however, fitted the yucca for the ungenial soil it is destined to occupy. The plant consists of a large tuft of rigid spear-pointed leaves, placed immediately upon the root, and sending up in the flowering season, a stalk bearing a cluster of lillaceous flowers as large as those of the common tulip of the gardens. The root bears more resemblance to the trunk of a tree, than to the roots of ordinary plants. It is two or three inches in diameter, descending undivided to a great depth below the surface, where it is impossible the moisture of the earth should ever be exhausted, and there terminates in numerous spreading branches. In some instances, the sand is blown from about the root, leaving several feet of it exposed, and supporting the dense leafy head, at some distance from the surface (Thwaites, 1905, 15: 232–233).

CULTIVATION

Soapweed is an attractive plant, maintaining green leaves throughout the winter. Its sharp spines make it a very effective border plant. It has been planted in cemeteries in western Kansas because its large white flowers often bloom on Memorial Day. Soapweed appears to be cultivated in the western portion of its range, but these "yucca ranches" are just the response of this plant to an open habitat caused by overgrazing or other disturbance. Several species of soapweed are used as ornamentals. A cultivar of our native *Y. glauca* is called "Rosea" for its flowers, which are tinted rose on the outside. Soapweed can be propagated by seed, offsets, and cuttings of stems, rhizomes, or roots. It likes good drainage, a sandy loam soil, and full exposure to the sun.

Poaceae
The Grasses

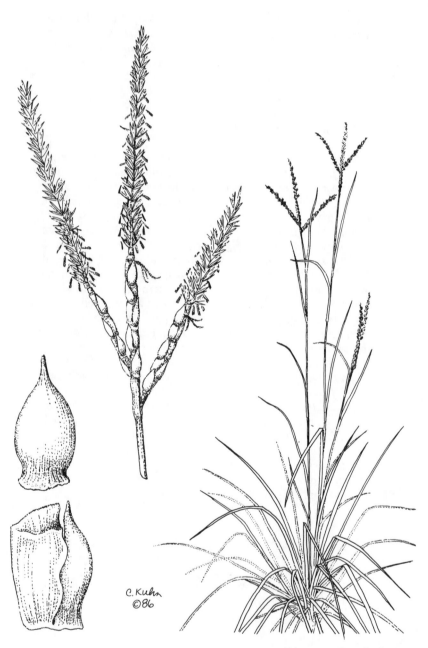

C. Kuhn
©86

Tripsacum dactyloides L.

The most important agricultural crops growing in the Prairie Bioregion are grasses: wheat, sorghum, and corn. In sheer number of species and individuals, wild grasses, with their edible seeds (referred to as grains), are the most important wild food plants of the prairies. However, most grasses have very small seeds tightly enclosed in inedible scales, which are hard to remove. Because of this, grass seeds generally were not a major food source.

This chapter will discuss grasses whose use has been recorded. The sources of information are studies of prehistoric archaeological sites and of recent Indian uses from the Southwest and the Great Basin, where grasses have greater importance because other sources of food are more limited. All of these grasses, or closely related species, grow in the prairies, and they were probably gathered for food during prehistoric times, especially in times of drought.

Alfred Whiting reported (1939, p. 18) that the seeds of wild grasses played an important role in the diet of the Hopi Indians of Arizona: "Though the main food supply comes from agriculture, there are a few wild plants which provide an important means of sustenance. Chief among these are several wild grasses." For the Gosiutes in Utah, R. V. Chamberlin reported in 1911 (p. 340): "While many kinds of plants fur-

nished seeds that were used, by far the greater portion come from the grasses and Chenopodiaceae." John Doebley reported (1984, pp. 61, 62) that 53 wild grasses were gathered in the Southwest for food. Of these, 74 percent were perennials. "Perennial grains were readily available to these people and they have the following advantages over annuals as a food source: (1) they produce seed more reliably during years of drought; (2) their roots and growing points are formed during the previous season and are ready to grow when the first spring rains come; (3) some species are able to develop mature seeds earlier in the growing season than annual species; and (4) in undisturbed habitats, perennials can be superior competitors, forming larger stands."

Because of these advantages and because they can reduce soil erosion with their tenacious root systems, perennial grasses may play an important role in the development of a sustainable agriculture. Research is being conducted by Wes Jackson and others at the Land Institute of Salina, Kansas, to develop a perennial grain crop (Jackson, 1985).

In the prairies, some wild grass seeds were prehistoric food sources. For five archaeological sites that were studied in the Middle Missouri River region in North and South Dakota, "it appears that much greater emphasis was placed on the native cultigens *Helianthus* and *Iva*, along with indigenous weedy plants and grasses," than was previously supposed (Nickel, 1977, p. 57). Wild grasses are also among the probable plant foods whose remains were found at the Two Deer site, near El Dorado, Kansas, which was occupied between 885 and 1060 A.D. (Adair, 1984, p. 195).

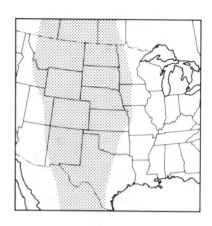

GENERAL DESCRIPTION

Annual or perennial herbs sometimes growing from runners or rhizomes, with erect, slender stems, sometimes branched. Leaves alternate, long and narrow, with pointed tips and extended bases that wrap around stems. Flowers in small clusters at tops of plants, often arranged on branches; each one with several small, leaflike structures, no petals, and dangling yellow anthers bearing pollen. Fruits small, dry, each containing 1 seed.

Bouteloua gracilis (H.B.K.) Griffiths

Blue Grama

The seeds of the blue grama were reported to have been eaten during prehistoric times by the White Mountain Apache (in Arizona) (Reagan, 1929, p. 155).

Elymus canadensis L.

Canada Wild Rye

The seeds of Canada wild rye were formerly gathered widely by the Gosiute (in Utah) for food (Chamberlin, 1911, p. 368).

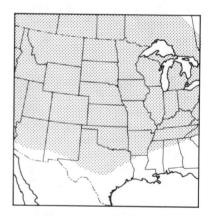

Festuca Species

Fescue

A species of fescue has been recognized in several maize-bearing Late Woodland sites in west-central Illinois, often representing over 20 percent of the identifiable seeds. It has tentatively been identified as *F. paradoxa* Desv. or *F. obtusa* Biehler (Asch and Asch, 1982, p. 18).

A related annual, cool-season grass, *F. octoflora* Walt., was used by the Navaho as a grain crop, which they planted along with their corn (Wyman and Harris, 1951, p. 16). The Navaho name for this grass means "afraid of the summer," because it ripened with the first warm weather. It also ripened before the corn, providing food until that crop was ready (ibid.).

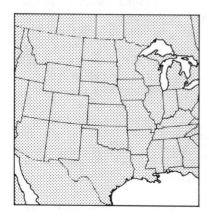

Koeleria pyramidata (Lam.) Beauv.

Junegrass

At Isleta pueblo in New Mexico, the seeds of the cool-season junegrass were important for food before the introduction of wheat. Meal prepared from the seeds was made into both bread and mush (Castetter, 1935, p. 22).

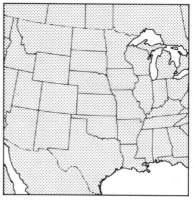

Muhlenbergia Species

Muhly

The seeds of some species of *Muhlenbergia* were threshed, winnowed, and ground into flour for bread by the Mescalero and Chiricahua Apache (Opler, 1936).

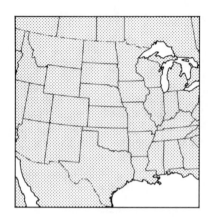

Oryzopsis hymenoides
(R. & S.) Ricker

Indian Ricegrass

Indian ricegrass was probably the most valuable wild grass utilized by Indians of the Southwest. It is a perennial that grows on deserts and plains throughout the western United States. It was used extensively because it is a widely dis-

tributed cool-season grass with large seeds that ripen in the early summer. Its abundance is partially due to its ability to thrive in disturbed habitats (Doebley, 1983, p. 59).

Indian ricegrass, also called Indian millet and sand bunchgrass, was used by the Indians at Zuni Pueblo (in New Mexico) westward to California (Castetter, 1935, p. 27) and perhaps as far north as Montana (Blankenship, 1905, p. 11). The Zuni declared that the seeds of this grass formed one of their staple foods before they had corn (Stevenson, 1915, p. 67). Edward Palmer reported (1871, p. 419) that: "This is a singular species of grass which is found growing wild in moist sandy spots in Nevada, Arizona, and New Mexico, and produces a small, black, nutritious seed, which is ground into flour and made into bread. It is held in high estimation by the Zuni Indians of New Mexico, who, when their farm crops fail, become wandering hunters after the seeds of this grass, which is abundant in their country. Parties are sometimes seen ten miles from their villages, on foot, carrying enormous loads for winter provision."

The Zuni also mixed ground Indian rice grass with cornmeal and water and formed it into balls or pats, which they would steam and eat (Stevenson, 1915, p. 67). The seeds are also eaten by the Hopi and Navaho. The Hopi have used

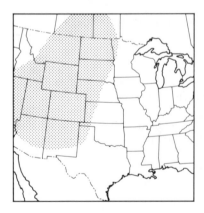

the seeds since ancient times and one of their clans is named for this plant (Castetter, 1935, p. 28). The Navaho name is "in-dit-lith-ee" (burnt off or burnt free) in reference to the adhering chaff, which is burned off the edible seeds (ibid.). Indian ricegrass seeds also served as food for the White Mountain Apache (in Arizona), the Gosiute (in Utah), and the Panamint or Koso (in southern California) (ibid.).

Panicum Species

Panic Grass

The seeds of panic grasses were used widely in the Southwest. However, their importance as a food source was less than other grass seeds because they dispersed quickly from the plant after they were ripe (Doebley, 1983, p. 60). The Hopi (in Arizona) used the

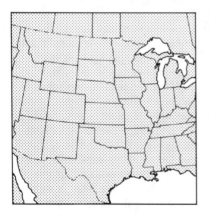

ground seeds of panic grasses, *P. capillare* L. and *P. obtusum* H. B. K., mixed with cornmeal for food (Castetter, 1935, pp. 28, 38). Other species, *P. obtusum* and *P. bulbosum*, were used by the Mescalero and Chiricahua Apache. After threshing and winnowing, the grain was ground into flour and baked for bread. This flour was also used to make gravy that was mixed with meat (Opler, 1936, p. 48). In the Sonoran Desert region, the native grain, *P. sonorum* Beal., was harvested, sown, and culturally selected as a prehistoric domesticated crop (Nabhan and De Wet, 1983, p. 65).

Phalaris caroliniana Walt.

Maygrass

The native range of maygrass extends northwest to the forest/prairie transition of Kansas and Missouri, but it is rare north of central Arkansas. Maygrass is an annual, whose common name tells when its seeds are ripe. It was used extensively by prehistoric Indians in the east-central United States. Maygrass is one of the three annual starchy-seeded species that dominate Middle and Late Woodland seed collections in west-central Illinois (where it does not grow in the wild today) (Asch and Asch, 1982, p. 13). The abundance of maygrass at these sites strongly suggests that it was a cultivated crop (ibid.). Maygrass has a life cycle similar to that of wheat (being a fall-planted, winter annual) and was probably an important late-spring-to-early-summer food source.

The archaeological range of maygrass is from New Mexico and Alabama, north to southwest Indiana (Cowan, 1978, p. 274; Bohrer, 1975, p. 199). A few maygrass

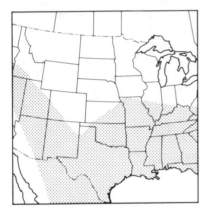

seeds, found at the Koster site along the Illinois River in southwest Illinois, have been dated to 4600 B.C. (Asch and Asch, 1982, p. 14). The best evidence that maygrass was utilized as a food comes from the analysis of prehistoric coprolites from Newt Kash Hollow and Salts Cave and Mammoth Cave in Kentucky. Of the 154 coprolites that have been examined, 27 percent contained varying amounts of maygrass grain (Cowan, 1978, p. 279).

At Tonto National Monument in Arizona, well west of the current range, about 60 cut heads of maygrass were found in the screenings from the household debris from one room. Since the seeds were missing, it was assumed that the heads had been processed (Bohrer, 1975, p. 199). Based on current archaeological and botanical evidence, it is not certain that maygrass was ever a true domesticate, like marsh elder (*Iva annua*) or sunflower (*Helianthus annuus*) (Cowans, 1978, p. 284). However, the early association of maygrass with a series of manipulated tropical and indigenous eastern North American cultigens indicates that it was at least an encouraged and protected plant, which grew in and was harvested from gardens (ibid.).

Sporobolus cryptandrus (Torr.) Gray

Sand Dropseed

Sand dropseed and other *Sporobolus* species were important food resources in the Southwest because they are relatively high yielding and have naked seeds (without adherent scales) that tend to remain attached to the plant after maturation (Doebley, 1983, p. 60). The Navaho reportedly made dumplings, rolls, and griddle cakes from sand dropseed flour, whereas the Hopi ground these seeds and mixed them with

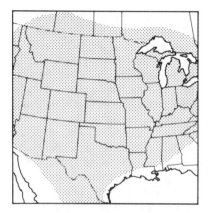

Tripsacum dactyloides L.

Eastern Gama Grass

Eastern gama grass is native to the tallgrass prairie and is related to corn. It has large seeds that are tightly closed in a tough outer hull. Melvin Gilmore (1931b, p. 91) found them in the remains of the Ozark Bluff-dwellers: "A puzzling fact is the presence of great quantities of remains of *Tripsacum dactyloides* L. in the Bluff-Dweller habitations. The seeds undoubtedly would be good for food, but they are so deeply imbedded in the hard, tough rachis that one cannot see how the Indians could have separated them. The hulls might possibly have been cracked by pounding in a mortar and then winnowed out."

cornmeal (Castetter, 1935, p. 28). Sand dropseed was threshed, winnowed, and ground into flour for bread by the Chiricahua and Mescalero Apache. They also boiled the seeds and ate them in porridge (Opler, 1936, p. 48).

At Clydes Cavern in east-central Utah, dropseed was found in 65 percent of the coprolites that were examined from the last five levels of cultural materials (Winter and Wylie, 1974, p. 309). Dropseed was also found in many of the coprolites from a prehistoric Lower Pecos site in southwest Texas. However, Glenna Williams-Dean (1978, p. 187) has suggested that because of the small quantity of seeds in each coprolite, they were probably accidentally ingested when an ungutted rodent or bird was eaten. The availability and productivity of dropseed species probably made them a food source wherever they grew.

It has also been suggested that eastern gama grass seeds might have been popped to render them edible. The popped kernels of *Tripsacum* species are almost indistinguishable from those of open-pollinated strawberry popcorn (Komarek, 1965, p. 216).

Eastern gama grass has been selected by the Land Institute in Salina, Kansas, as a native prairie plant to develop as a perennial grain crop. (Updates on their research on perennial grain crops and other experiments are available in their informative journal, *The Land Report.*) The following virtues make eastern gama grass a good candidate for a productive

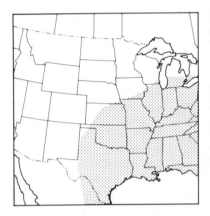

perennial crop (Jackson and Bender, 1978, p. 6; Jackson, 1986, personal communication): (1) it has already been extensively studied, particularly by those interested in the evolution of corn; (2) it is native to the corn belt of the United States; (3) it can easily be propagated vegetatively for breeding; (4) it can readily be manipulated genetically; (5) the species has natural immunity against grasshoppers, corn ear worm, and the European corn borer; (6) it is able to fix small amounts of atmospheric nitrogen; and (7) it has a protein content of over 27 percent and a carbohydrate content of over 51 percent.

Zizania aquatica L.

Wild Rice

Wild rice is a plant of shallow lakes and marshes, whose western distribution extended into the prairies. It is also called water rice, water oats, and Indian rice and had the following Indian names: Dakota, "psin" (see prairie turnip, *Psoralea esculenta*, for the relationship between the meaning and origin of the two Dakota names); Omaha and Ponca, "si-waninda"; and Winnebago, "sin" (Gilmore, 1977, p. 15). Wild rice has been used extensively by the Indians of the Great Lakes Bioregion. In 1671, a French Jesuit priest reported (in present-day Minnesota) that the Dakota "content themselves with a kind of marsh rye that we call *folle avoine*, which the prairies supply spontaneously (Jenks, 1900, p. 1046). For the Dakota, who moved westward to the Nebraska and South Dakota area, wild rice was such an important and prized food that they gave its name to their moon that corresponds to our month of September: "psin-hna-ketu-wi" means "the moon to lay rice up to dry" (Gilmore, 1977, p. 360).

The Omaha ate wild rice, which they found in oxbow lakes along the Missouri River. Edwin James, botanist on the Long Expedition, reported in 1819 that about two miles from the Omaha village is a large pond "which is filled with luxuriant aquatic plants, amongst which the zizania and nelumbium, are particularly worthy of note both for their beauty and importance for economical pur-

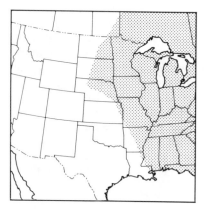

poses" (Thwaites, 1905, 14: 289). Around 1910, an Omaha man said, when referring to wild rice: "We used to use this, but we don't now. Before the white man came we had to eat whatever we could get of just what grew in the country then" (Gilmore, 1913a, p. 328).

Wild rice is a nutritious grain, high in calories and protein (Watt and Merrill, 1963, p. 67). Its native range extended into the Sand Hills of Nebraska. Melvin Gilmore (1977, p. 15) stated while working for the Nebraska State Historical Society: "This cereal was an important part of the dietary of the tribes of Nebraska, but not in so great a degree as with the tribes of the lake regions toward the northeast. It would seem worthwhile to raise wild rice in any lakes and marshy flood plains in our State not otherwise productive, and so add to our food resources. From trial I can say that it is very palatable and nutritious and, to my taste, the most desirable cereal we have."

Today, wild rice is grown in manmade paddies in Minnesota and other northern states. It is marketed as a gourmet food, valued mainly for its unique flavor but also for its high protein content.

Other Edible Prairie Plants

This section includes species that were less important as prairie food sources. Extra caution should be taken when using them because they are less well known than other edible prairie plants.

Acorus calamus L.

CALAMUS (SWEET FLAG)
Araceae
Arum Family

Calamus is a grasslike plant with sword-shaped leaves and thick, cylindrical spikes of tiny, brown flowers. It grows in marshy or wet habitats in the eastern Prairie Bioregion. It was prized by the Indians of the prairies for its medicinal, ritualistic, and food uses. The Pawnee name is "kahtsha itu" (medicine lying in the water) (Gilmore, 1977, p. 18). The Osage call it "peze boao'ka" (flat herb) and the Lakota call it "sinkpe tawote" (muskrat food) and specifically call the root "sunkace" (dog penis), probably in reference to the shape of its flower stalk (Munson, 1981, p. 231).

The Osage chew the root for its flavor and the Lakota eat the leaves, stalks, and the roots (Munson, 1981, p. 231). The young, tender leaves add a fresh, clean flavor to spring salads. The Omaha were reported to eat the root boiled, although this may have been more for medicinal use (Dorsey, 1882, p. 308; Howard, 1965, p. 43). Cala-

mus rhizomes were also reported to be "conspicuously abundant" in the remains of the Ozark Bluff-dwellers, indicating medicinal and ritualistic use (Gilmore, 1931b, p. 95). The Pawnee, Lakota, and other tribes held this plant in high esteem: "When a hunting party came to a place where the calamus grew the young men gathered the green blades and braided them

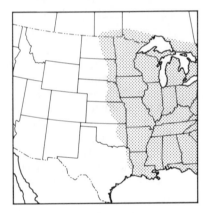

into garlands, which they wore round the neck for their pleasant odor. It was one of the plants to which mystic powers were ascribed. The blades were used also ceremonially for garlands. In the mystery ceremonies of the Pawnee are songs about the calamus" (Gilmore, 1977, p. 18). The western distribution of calamus in Nebraska corresponds closely to old Pawnee village sites, suggesting that it was planted in these locations (ibid., p. 8).

Agastache foeniculum (Pursh) O. Ktze.

LAVENDER HYSSOP
Lamiaceae
Mint Family

This robust perennial has opposite, egg-shaped leaves with toothed margins and small, blue to violet flowers in whorls around the upper stems. It is found in upland woods and on prairies in the northern Prairie Bioregion. The Indians of the Missouri River region were reported to have used the leaves for tea and also as a flavoring in cooking (Gilmore, 1977, p. 61). The Lakota name for this plant is "wahpe' yata'pi" (leaf that is chewed) (Rogers, 1980a, p. 76) and the Cheyenne name is "mo e'-emohk' shin" (elk mint) (Grinnell, 1962, p. 186). Flowers of this plant were frequently included in Cree medicine bundles (Johnston, 1970, p. 318). Lavender hyssop can easily be propagated by seed sown

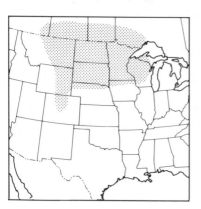

in the spring or fall, cuttings, or root division (Foster, 1984, p. 61). It thrives in a sandy, well-drained rich and loamy soil and prefers full sun, but will tolerate some shade.

Antennaria Species

PUSSY-TOES
Asteraceae
Sunflower Family

These perennial herbs are covered with whitish hairs. Leaves are spoon-shaped and mostly at the base. Flower heads are compact and fuzzy, at the tops of stalks.

The name pussy-toes refers to the shape of these heads, which later contain small, dry fruits that are dispersed in late spring. The leaves of *Antennaria microphylla* Rydb. (formerly *A. rosea*), in which some of the scales around the flower heads are pink, were chewed by Blackfoot children for their flavor (Johnston, 1970, p. 320). Field pussy-toes, *A. neglecta* Greene, found on waste ground and shallow soils of prairies has been reported to be chewed like gum (Phillips, 1979, p. 50).

Artemisia Species

SAGE
Asteraceae
Sunflower Family

White sage, *Artemisia ludoviciana* Nutt., is a hairy, perennial herb, with alternate, elliptic leaves, entire or irregularly lobed.

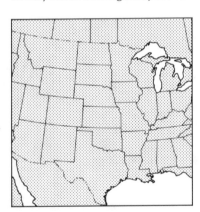

Flower heads are small and greenish, in tight clusters among the leaves at tops of stems. It grows on prairies throughout the Prairie Bioregion. The Lakota called it "peji'ho'ta ape'blaskaska" (gray herb with a flat leaf) (Rogers, 1980b, p. 36). They burned this species as an incense to induce the presence of good influences (Gilmore, 1977, p. 83). In the Great Basin and Southwest, the tiny seeds of several sage species were reported to be used as food by the Indians (Yanovsky, 1936, p. 59). The seeds of big sagebrush (*A. tridentata* Nutt.) are considered to be the best tasting; they can be eaten raw or dried and pounded into meal (Rogers, 1980a, p. 49). This species is also hairy and aromatic, but is bigger, with alternate leaves divided into 3 teeth near the tips. Sagebrush has a more western distribution.

Astragalus canadensis L.

CANADIAN MILKVETCH
Fabaceae
Bean Family

Canadian milkvetch can be found growing on riverbanks or moist prairies. It was gathered by the Blackfoot, who "dug them in spring for eating, preparing them

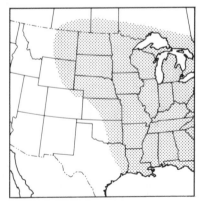

for eating by boiling" (McClintock, 1909, p. 278). It looks similar to some closely related poisonous locoweeds, so its use is not recommended unless positive identification can be made. This species differs from *A. crassicarpus*, described above, in having yellowish or greenish-white flowers. It grows in moist prairies and open woods.

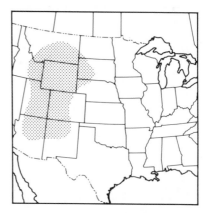

Calochortus gunnisonii S. Wats.

SEGO LILY
Liliaceae
Lily Family

This onion-shaped perennial is found on dry prairies and open (often rocky) coniferous and deciduous woods in the northwestern part of the Prairie Bioregion. It has 2–4 narrow, linear leaves. The white to slightly purple flower, usually singular, is quite showy in July and August with its six petals. The long, sweet-tasting bulbs of the sego lily (also commonly called the mariposa lily) were used by the Cheyenne, who gathered, dried, and stored them for winter use. The dried bulbs were pounded fine and boiled. According to George Bird Grinnell (1962, p. 172), this "makes a sweet porridge or mush. When cooked fresh, these bulbs are very tender and likely to fall to pieces."

There is a second sego lily species, *C. nuttallii* T. & G., found in the same area, that is quite similar in appearance and is also edible. The Dakota (Sioux) name for both sego lilies is "psin tanka" (big onion) (Rogers, 1980a, p. 25). These tasty roots can also be eaten raw. Care should be taken to not overharvest this plant.

Celtis reticulata Torr.

NETLEAF HACKBERRY
Ulmaceae
Elm Family

This shrub-sized hackberry has gray bark with corky ridges and alternate, egg-shaped leaves, with entire margins and lower surfaces yellow-green with conspicuous veins. The fruits are fleshy, round, and reddish. It is most common on dry limestone hills and ravine banks in the southern portion of the Prairie Bioregion. It has edible fruits, as do all of the hackberry species.

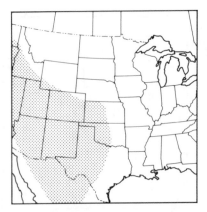

The fruits of the hackberry were used by the Osage in hackberry cakes, which were stored in caches over the winter (Matthews, 1961, p. 443). The Chiricahua and Mescalero Apache were reported in 1936 to use the fruit of the net-leaf hackberry, after it ripened in September, either fresh or ground and shaped into cakes, which would be stored for winter use (Opler, 1936, p. 46). Sometimes they also made jelly from the fruit (ibid.). It is not clear whether the large hard stone in the middle of the fruit was removed before the fruits were pounded. The Dakota were known to pound up the hard stone and fruit of the hackberry (*C. occidentalis* L.) to use as seasoning for meat (Gilmore, 1977, p. 24). This species is a tree with toothed leaves and fruits that turn dark purple to brown when ripe in the fall and are sweet tasting.

Grindelia squarrosa (Pursh) Dun.

CURLY-TOP GUMWEED
Asteraceae
Sunflower Family

Curly-top gumweed is a biennial or short-lived perennial found in waste places and pastures. It has alternate leaves, oblong, with toothed margins. Sticky, resinous sap covers the leaves and the curly, modified leaves under the yellow, daisylike flower heads. These sticky flower heads can be used as a chewing gum substitute, and the young leaves make a pleasantly aromatic, slightly bitter tea (Moore, 1979, p. 82).

The Pawnee call curly-top gumweed "bakskitits" (sticky head); the Dakota name is "pte-ichi-yuha" (curly buffalo); and the Omaha and Ponca name is "pezhe-wasek" (strong herb) (Gilmore, 1977, p. 81). The plant was used for medicinal purposes by these tribes (ibid.).

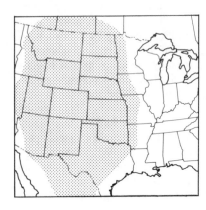

Lathyrus polymorphus Nutt.

HOARY PEAVINE
Fabaceae
Bean Family

This perennial herb has alternate, compound leaves, each with an obvious bristle at the tip. Flowers are typical of the bean family, fragrant, with upper petals purple and lower ones sometimes white. Fruits are dry and narrow, with round, brown, and smooth seeds. It grows on prairie hillsides, in stream valleys, and in open woods.

The seeds of the hoary peavine were a minor food source of the Omaha and Ponca. When they went hunting in the Sand Hills (in Nebraska), "the children sometimes gathered the pods, which they roasted and ate in sport" (Gilmore, 1977, p. 46). The Omaha and Ponca name for this plant is "hinbthi-si-tanga" (large

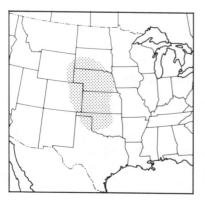

seeded bean) (ibid.). The entire pods of this plant were used as food by the Acoma, Laguna, and Cochiti pueblo people of the Southwest (Castetter, 1935, p. 32). The leaves of some species of Lathyrus are poisonous, so care should be taken to identify and carefully test this plant before it is consumed in quantity (Fernald et al., 1958, p. 252).

Lespedeza capitata Michx.

ROUND-HEAD LESPEDEZA
Fabaceae
Bean Family

This common prairie plant is a perennial herb, with alternate leaves, divided into 3 elliptic leaflets. Flowers are small, white, and arranged in round clusters among the upper leaves. The clusters of dry, brown fruits that develop in the fall resemble rabbit feet. It also grows in old fields and along roadsides.

The leaves of round-head lespedeza were boiled in water by the Comanche for tea (Carlson and Jones, 1939, p. 531). The Pawnee name for this plant is "parus-as" (rabbit foot) and the Omaha and Ponca name is "te-hunton-hi nuga" (male buffalo bellow plant) (Gilmore, 1977, p. 45). The latter tribes used it for medicinal purposes and found it growing on the hills of the loess plain, whereas the "female buffalo bellow plant," leadplant (Amorpha canescens),

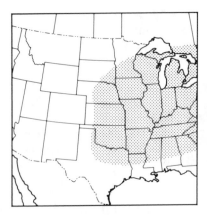 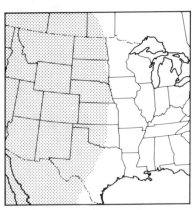

which was used similarly, was gathered from the sandy loam soils of valleys (ibid.). Both of these plants bloom when buffalo bulls bellow during their rutting season.

Linum perenne
L. var. *lewisii*

BLUE FLAX
Linaceae
Flax Family

Blue flax is a perennial herb with narrow, alternate leaves. Flowers have 5 separate, blue petals. Fruits are dry and egg-shaped and separate into 10 segments, to release flattened and slimy seeds. It can be found in prairies and open, rocky hillsides in the western portion of the Prairie Bioregion.

The seeds of blue flax, also called perennial flax and prairie flax, were reported to be gathered and used in cooking by the Indi-

ans of the Upper Missouri River region because they added an agreeable flavor to cooked foods and were highly nutritious (Gilmore, 1977, p. 46). Captain Meriwether Lewis of the Lewis and Clark expedition reported on July 18, 1805 (while in present-day Montana), that he had been seeing blue flax for several days and thought that if it had good fibers in its stems, it might make a good crop plant (for fiber and linen) because of its perennial nature (Thwaites, 1904, 2: 244).

Mentzelia albicaulis
(Hook.) T. & G.

MENTZELIA
Loacaceae
Stickleaf Family

Mentzelia is an annual plant with white stems and alternate, linear to lance-shaped leaves. The flowers have five yellow to orange petals. The plant is found on dry, gravelly, and sandy soils in the western Prairie Bioregion. The Hopi, in the Southwest, parch and then grind the small, oily seeds into a fine, sweet meal and eat it in pinches (Castetter, 1935, p. 34).

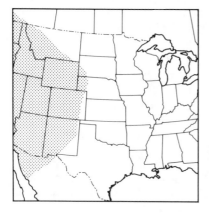

Nelumbo lutea
(Willd.) Pers.

AMERICAN LOTUS
Nelumbonaceae
Lotus Family

This aquatic plant, which is found in oxbow lakes and around the shallow edges of ponds, grows from a banana-shaped rhizome. The leaves are round and usually above the water. The showy flowers, with many pale yellow petals, are also above water. The fruits are dry, round, and nutlike, embedded in round, flat-topped structures. American lotus grows in lakes and ponds in the eastern part of the Prairie Bioregion.

American lotus was highly prized and considered to be invested with mystic powers by the Omaha and other Missouri River tribes (Gilmore, 1977, p. 27). The hard seeds were cracked and freed of their shells and used with meat for making soup. The tubers, after being peeled, were cut up and cooked with meat or with hominy and contributed a delicious flavor, unlike any other (ibid.).

The Omaha used this plant extensively. Edwin James, botanist for the Long Expedition, reported in 1819 that the American lotus was growing in a pond about two miles from the Omaha village (Thwaites, 1905, 14: 289). Dr. F. V. Hayden stated in his "Botany Report" (1859, p. 730) that the American lotus was found on the "lower portion of the valley of the Platte (River) in the broad, wet bottoms about Omaha City" and that "it is now quite rare on account of the great use of both roots and seeds for food, by the Omaha, Otoe, and Pawnee Indians." Owen Dorsey reported (1882, pp. 308–309) the following description of the American lotus by an Omaha:

The "tethawe" is the root of an aquatic plant, which is not very abundant. It has a leaf like that of a lily, but about two feet in diameter, lying on the surface of the water. The stalk comes up through the middle of the leaf, and projects about two feet above the water. On top is a seed-pod. The seed are elliptical, almost shaped like bullets, and they are black and very hard. When the ice is firm or the water shallow, the Indians go for the seed, which they parch by a fire, and beat open, then eat. They also eat the roots. If they wish to keep them for a long time, they cut off the roots in pieces about six inches long, and dry them; if not, they boil them.

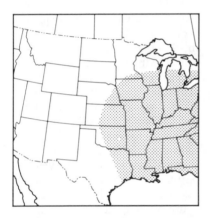

The American lotus is quite nutritious, containing significant amounts of starch and protein (Yanovsky and Kingsbury, 1938, p. 655). If the seeds are harvested just as they mature, they are relatively easy to remove from their shells and taste like nuts. After they have fully matured and dried, the shells and nuts are both extremely hard.

This is not a common plant and should be encouraged in shallow ponds and lakes with clean water for its beautiful flowers and as a food source. In New York, Connecticut, Delaware, and Ohio, there is evidence that it may have been planted by the Indians (Newberry, 1888, p. 39). It is quite likely that the Indians propagated this plant to the limits of its range in the Prairie Bioregion.

Oenothera biennis L.

COMMON EVENING PRIMROSE
Onagraceae
Evening Primrose Family

The common evening primrose is a biennial herb that grows from a taproot. The basal leaves are in rosettes, often red-spotted and lobed; the stem leaves are alternate and lance-shaped. Flowers have 4 separate, white petals. The fruits are dry and cylindrical. It can be found in open woods and along streams and roadsides.

This species at one time was a popular vegetable that was introduced and cultivated in Europe for its root. The roots are harvested in the fall or early spring and at their best, taste similar to a sweet parsnip. Roots that I harvested in the early fall and roasted in ashes were quite tasty. The basal leaves can be used in the fall and spring as a pot herb, although they sometimes have a bitter taste. It is best to cook all parts of the plant because the bitter taste can be par-

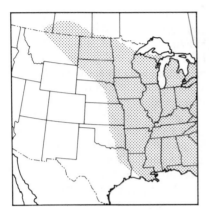

It was reported that the entire plant was eaten by the Gosiute in Utah, where its roots parasitize the sagebrush plant (Chamberlin, 1911, p. 361). It is also a parasite on roots of giant ragweed (*Ambrosia trifida*), curly-top gumweed (*Grindelia squarrosa*), cocklebur (*Xanthium strumarium* L.), and other members of the Sunflower Family (Steyermark, 1981, p.1374).

tially removed through changing the water and cooking again. Some roots also have a woody string in their core, which is probably due to age. It has also been suggested that other related evening primroses are edible, and none of them have been reported to be poisonous (Harrington, 1967, p. 84).

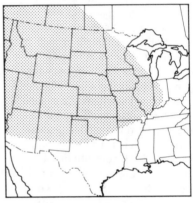

Orobanche ludoviciana Nutt.

BROOMRAPE
Orobanchaceae
Broomrape Family

Broomrape or cancer-root is a parasitic plant of dry, sandy, upland prairies. It consists of stout stems with scalelike leaves, white to pink or yellow (lacking chlorophyll). The flowers are small, tubular with 5 lobes, and off-white to purple.

Edward Palmer (1878, p. 605) reported that broomrape was a food source of the Paiutes in Utah: "All the plant except the bloom grows under ground, and consequently is nearly all very white and succulent. The Pah-Utes consume great numbers of them in summer while on their hunting excursions after rabbits. Being succulent they answer for food and drink on these sandy plains, and, indeed, are often called 'sand food.'"

Polygonum Species

KNOTWEED
Polygonaceae
Buckwheat Family

The Sioux were reported to eat the young shoots of swamp smartweed in the spring as a relish (Blankenship, 1905, p. 18). This species, *Polygonum amphibium* L. var. *emersum* Michx. (formerly *P. coccineum*), is a perennial herb with alternate, lance-shaped leaves and small, pink flowers in elongated clusters at tops of stalks. It grows in wet areas and also along railroads.

Knotweed seeds were a prehistoric food source and are frequently found in archaeological remains, such as the Yeo site, a Kansas City Hopewellian village that was occupied from 635 to 870 A.D. (O'Brien, 1982, p. 37). Knotweed, *P. erectum* L., differs from the swamp smartweed in being an annual with broader leaves and

greenish flowers in small groups among the leaves. The fruits are brown and 3-angled. It is usually found growing in drier places and produces a starchy seed, which, along with goosefoot (*Chenopodium*) and maygrass (*Phalaris caroliniana*), dominates the Middle and Late Woodland seed assemblages found in west-central Illinois (Asch and Asch, 1982, p. 14). It is believed that this species was probably a prehistoric cultivated crop (ibid.).

Potentilla Species

CINQUEFOIL
Rosaceae
Rose Family

The several species of cinquefoil having leaves with 5 fingerlike segments and flowers with 5 separate, yellow petals are common plants of some prairies, meadows, and waste grounds. The early

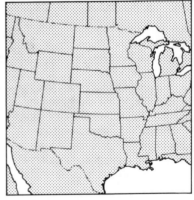

leaves can be dried and used for a good-tasting, golden-colored tea that is high in calcium (Phillips, 1979, p. 67).

Pycnanthemum Species

MOUNTAIN MINT
Lamiaceae
Mint Family

Mountain mint is not an appropriate name for these plants that grow in prairies and woodlands of the eastern portion of the Prairie Bioregion. They have square stems up to 12 dm (48 in) tall, with opposite, egg-shaped to lance-shaped leaves. The flowers are small, tubular, white to pale lavender, and arranged in round, branching clusters at tops of stalks. The leaves have a noticeable, strong mint smell when bruised and make a tasty tea.

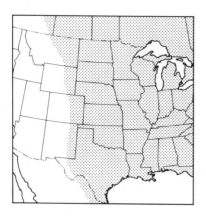

Sambucus canadensis L.

ELDERBERRY
Rubiaceae
Madder Family

Elderberry is a shrub with opposite, compound leaves; leaflets have toothed margins. The small flowers are in flat-topped, round clusters at the ends of branches; petals are creamy white, fused at

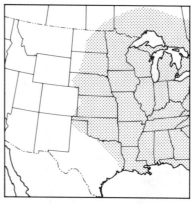

the bases and separating into 5 lobes. The fruits are fleshy, round, and dark purple. This plant favors moist areas in woods, roadside ditches, and stream banks.

The foliage and roots of elderberry are poisonous and should be avoided. However, the fruits can be gathered when ripe in the late summer or fall. They can be used raw (although somewhat bitter), cooked, or dried, and the juice is often made into jelly or wine. The Dakota, Pawnee, Omaha, and Ponca were reported to eat elderberry fruit (Gilmore, 1977, p. 63). These fruits are a good source of vitamin C (Watt and Merrill, 1963, p. 30). The Ponca used the blossoms for tea by dipping them into hot water (Howard, 1965, p. 45). Also, the mildly almond-flavored flowers can be dipped in batter and fried as fritters.

Silphium perfoliatum L.

CUP PLANT
Asteraceae
Sunflower Family

This perennial has square stems and leaves that are mostly opposite, egg-shaped, toothed, and with cuplike bases that hold water. Flower heads are large, with yellow ray florets, and numerous on each plant. Cup plant is found in wooded areas, moist ground, and roadside ditches in the eastern portion of the Prairie Bio-

region. Its young leaves can be cooked in the spring as an acceptable green.

Solidago Species

GOLDENROD
Asteraceae
Sunflower Family

Goldenrods are attractive prairie and roadside plants, most noticeable when they bloom in late summer and fall. They are perennial herbs with alternate leaves, lance-shaped to egg-shaped, usually toothed. Flower heads are small and yellow, arranged in conspicuous spikes, flat-topped clusters, or curving, plumelike branches. One species, *Solidago missouriensis* Nutt., has been reported to be a minor food source in the Southwest, where the Hopi ate its young leaves with salt (Castetter, 1935, p. 52). This species

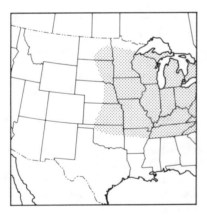 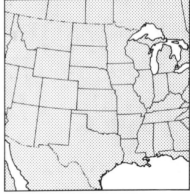

has narrow, toothed leaves and flower heads in curving branches. In the Great Basin, the Gosiute occasionally gathered the seeds of *Solidago* species for food (Chamberlin, 1911, p. 382).

Thelesperma magapotamicum (Spreng.) O. Ktze.

GREENTHREAD
Asteraceae
Sunflower Family

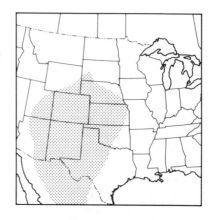

Greenthread is a perennial herb with opposite leaves, divided into very narrow segments. Flower heads, at the tops of stalks, consist of round clusters of tiny yellow florets (disks), and usually no ray florets. It grows in open sites in the southern portion of the Prairie Bioregion.

The Pueblo Indians have used greenthread as a tea plant for cen-turies (Moore, 1979, p. 66). The entire plant can be dried, either before or after it blooms, and stored for later use. Tea can be made by steeping the plant for a few minutes or by pouring boiling water over the leaves to leach out their flavor. The color of the tea is greenish-yellow to dark yellow-red, and its taste has been described as "delicious—we think the best of the native teas" (Harrington, 1979, p. 369).

Glossary

Aboriginal. Indigenous, native, or being the first of its kind in a region

Accrescent. Enlarging with age

Achene. A dry, indehiscent, 1-seeded fruit

Alkali. A salt or mixture of salts found in some soils

Alluvium. Material deposited by running water, such as sand or gravel in a river valley

Alternate. Placed singly one above the other on the axis

Annual. Living one year

Anthelmintic. An agent that expels intestinal, parasitic worms

Anther. Pollen-bearing part of a stamen

Antiquity. Ancient times or relating to the culture of ancient times

Astringent. Puckery; able to draw together the soft organic tissues

Bioregion. A geographical province defined by its unique life forms

Blade. The expanded portion of a leaf

Bract. A modified, reduced leaf

Bristle. A stiff hair

Bulb. An underground bud with thick, fleshy scales

Bulblet. A small bulb

Cache. A hiding place for concealing and preserving food

Cartographer. A person who makes maps

Cathartic. An agent that clears the bowels; a laxative

Compound. Made up of two or more parts

Confection. A sweet fruit, nut, or medicinal preserve

Coprolite. Preserved or fossilized feces

Copse or Coppice. A thicket or grove of small trees

Corm. The fleshy, bulblike base of a stem, usually underground

Crown. The portion of a stem at the surface of the ground

Cultigen. A cultivated variety for which a wild ancestor is unknown

Cultivar. Variety originating and persisting under cultivation

Declivity. Descending slope or inclination

Disk floret. A tubular flower in the central part of the flower head of plants in the Sunflower Family

Dispersal. The process of spreading organisms from one place to another

Ecologist. A person who studies the relationship of organisms with their environment

Ecosystem. The complex of a community and its environment forming a functioning whole in nature

Emetic. An agent that produces vomiting

Entire. Whole; with a continuous margin

Esculent. Edible; food to eat

Ethnobotany. The study of the uses of plants by a culture

Ethnography. The systematic recording of human cultures

Feral. Wild; not domesticated or cultivated

Fibrovascular. Having or consisting of fibers and conducting cells

Forb. An herb other than a grass

Glade. An open space surrounded by woods

Gland. A secretory organ

Glandular. Having glands

Head. A short, dense cluster of sessile or nearly sessile flowers

Herb. A plant lacking persistent woody parts aboveground

Herbaceous. Having the character of an herb

Hull. The outer covering of a seed or fruit

Indigenous. Produced, growing, or living naturally in a particular region or environment

Inflorescence. The flowering portion of a plant

Inulin. A carbohydrate (sugar) found in roots of plants of the Sunflower Family

Kernel. The inner soft part of a seed, fruit stone, or nut

Mucilaginous. Slimy

Mush. A cooked grain, usually cornmeal, either whole or ground

Opposite. Leaves arranged two at each node, on opposite sides of the stem

Palatable. Agreeable to the palate or taste

Parasite. A species living on or benefiting directly from another

Parfleche. An article made of rawhide, with the hair removed

Pemmican. An Indian food made of pounded, dried meat, berries, and fat

Perennial. Living several years

Petal. One division of the corolla in flowers

Pinnately compound. Leaf with leaflets arranged on both sides of the leaf axis

Pomade. A perfumed ointment, often used for hair or scalp

Pot herb. A green plant used for food or seasoning (usually cooked)

Prickle. A small, sharp outgrowth on the surface of a plant

Propagate. To increase the number of plants by sexual or asexual methods

Prostrate. Lying flat on the ground

Purgative. A strong laxative

Ray floret. A straplike marginal flower on the flower head of plants in the Sunflower Family

Restoration. The act of revegetating land, such as returning a parcel of land to native prairie species

Rhizome. An underground stem, usually lateral and rooting at the nodes

Root. An underground organ that anchors the plant and absorbs water and nutrients

Rootlets. Small roots

Rosette. A cluster of organs, usually leaves, arranged in a compact circle

Ruderal. Growing in waste ground or where man has disturbed the vegetation

Rut. An annual, recurring period of sexual excitement; the breeding season

Scale. A thin, dry, flattened organ

Saline. Containing salt

Scarify. To cut or soften the wall of a hard seed to hasten germination

Sepal. One division of the calyx, occurring below the petals in flowers

Slough. A wet, perhaps muddy and marshy area; swamp

Spike. An elongated inflorescence with stalkless flowers

Spine. A sharp, rigid outgrowth, usually from the wood of the stem

Stamen. The pollen-bearing organ of the plant

Stratify. To store in a cold, moist environment to promote germination

Style. The usually elongated part of the pistil between the ovary and the area that receives pollen

Succulent. Fleshy, juicy

Tannin. A soluble, astringent, complex phenolic substance of plant origin used in tanning, dying, ink, and medicine

Taproot. The primary descending root

Thresh. To separate the seed from the harvested plant

Tonic. A substance that increases strength and tone

Tuber. A thick short branch, usually subterranean, with numerous buds

Umbel. A flat-topped group of flowers on stalks arising from one central point

Voyageur. A man, usually French, who was employed to transport goods and men to and from remote stations in the Northwest

Wallow. A depression formed by the rolling of animals in the dirt

Wing. A thin, membranaceous extension of an organ; the lateral petal in flowers of the Bean Family

Winnow. To separate the grain from the chaff and dirt, usually by tossing in the air

Literature Cited

Adair, Mary J. 1984. Prehistoric Cultivation in the Central Plains: Its Development and Importance. Ph.D. dissertation, Department of Anthropology, University of Kansas.

Agogino, George, and Sherwin Feinhandler. 1957. Amaranth Seeds from a San Jose Site in New Mexico. *Texas Journal of Science* 9 (1): 154–156.

Arnason, Thor, Richard J. Herbda, and Timothy Johns. 1981. Use of Plants for Food and Medicine by Native Peoples of Eastern Canada. *Canadian Journal of Botany* 59 (11): 2189–2325.

Asch, David L., and Nancy B. Asch. 1977. Chenopod as Cultigen: A Re-evaluation of some Prehistoric Collections from Eastern North America. *Midcontinental Journal of Archeology* 2 (1): 3–45.

———. 1978. The Economic Potential of *Iva annua* and Its Prehistoric Importance in the Lower Illinois Valley. In: *The Nature and Status of Ethnobotany.* Edited by Richard I. Ford. Anthropological Paper 67, Museum of Anthropology, University of Michigan.

———. 1982. A Chronology for the Development of Prehistoric Horticulture in Westcentral Illinois. *Center for American Archaeology Archeobotanical Laboratory Report* 46.

———. 1985. Prehistoric Plant Cultivation in West-Central Illinois. In: *Prehistoric Food Production in North America.* Edited by Richard I. Ford. Anthropological Paper 75, Museum of Anthropology, University of Michigan.

Asch, Nancy B., and David L. Asch. 1980. The Dickson Camp and Pond Sites: Middle Woodland Archaeobotany in Illinois. In: *Dickson Camp and Pond Sites.* Edited by Anne-Marie Cantrell. Springfield: Illinois State Museum, pp. 152–160.

Bailey, Liberty Hyde, and Ethel Zoe Bailey. 1976. *Hortus Third: A Concise Dictionary of Plants Cultivated in the United States and Canada.* Revision by L. H. Bailey Hortorium Staff. New York: Macmillan.

Bailey, Ralph. 1962. *The Self-pronouncing Dictionary of Plant Names.* Garden City, N.Y.: American Garden Guild.

Bare, Janét. 1979. *Wildflowers and Weeds of Kansas.* Lawrence: Regents Press of Kansas.

Beardsley, Gretchen. 1939. The Groundnut as Used by the Indians of Eastern North America. *Michigan Academy of Science, Arts, and Letters* Paper 25: 507–515.

Bemis, W. P., L. D. Curtis, C. W. Weber, and J. Berry. 1978. The Feral Buffalo Gourd, *Cucurbita foetidissima. Economic Botany* 32: 87–95.

Benn, David W. 1974. Seed Analysis and Its Implications for an Initial Middle Missouri Site in South Dakota. *Plains Anthropologist* 19 (63): 55–72.

Black, M. Jean. 1978. Plant Dispersal by Native North Americans in the Canadian Subarctic. In: *The Nature and Status of Ethnobotany.* Edited by Richard I. Ford. Anthropological Paper 67, Museum of Anthropology, University of Michigan, pp. 255–262.

Blankenship, J. W. 1905. Native Economic Plants of Montana. *Montana Agricultural Experiment Station Bulletin* 56.

Bohrer, Vorsila L. 1975. The Prehistoric and Historic Role of Cool-Season Grasses in the Southwest. *Economic Botany* 29 (3): 199–207.

Bryant, Vaughn M., Jr. 1974. Prehistoric Diet in Southwest Texas: The Coprolite Evidence. *American Antiquities* 39 (3): 407–420.

Bye, Robert A., Jr. 1981. Quelites—Ethnoecology of Edible Greens—Past, Present, and Future. *Journal of Ethnobiology* 1 (1): 108–123.

Carlson, Gustav G., and Volney H. Jones. 1939. Some Notes on Uses of Plants by the Comanche Indians. *Michigan Academy of Science, Arts, and Letters* Paper 25: 517–542.

Castetter, E. E. 1935. Uncultivated Native Plants Used as Sources of Food. *University of New Mexico Ethnobiological Studies* 4: (1) 7–62.

Catlin, George. 1973 (1844). *North American Indians.* Vols. 1, 2. New York: Dover Publications.

Chamberlin, Ralph V. 1911. The Ethno-Botany of the Gosiute Indians of Utah. *American Anthropological Association Memoirs* 2 (5): 329–405.

Cole, John C. 1979. *Amaranth: From the Past, for the Future.* Emmaus, Pa.: Rodale Press.

Cowan, Wesley C. 1978. The Prehistoric Use and Distribution of Maygrass in Eastern North America: Cultural and Photogeographical Implications. In: *The Nature and Status of Ethnobotany.* Edited by Richard Ford. Anthropological Paper 67. Museum of Anthropology, University of Michigan, pp. 255–262.

Crane, Cathy J. 1982. Plant Utilization at Spoonbill, An Early Caddo Site in Northeast Texas. *Midcontinental Journal of Archaeology* 7 (1): 82–97.

Denig, Edwin T. 1930. Indian Tribes of the Upper Missouri— the Assiniboin. *Smithsonian Institution, Bureau of American Ethnology, 46th Annual Report.*

Dittmer, Howard J., and Burney P. Talley. 1964. Gross Morphology of Tap Roots of Desert Cucurbits. *Botanical Gazette* 125 (2): 121–126.

Doebley, John F. 1981. Plant Remains Recovered by Floatation From Trash at Salmon Ruin, New Mexico. *Kiva* 46 (3): 169–187.

———. 1984. "Seeds" of Wild Grasses: A Major Food of

Southwestern Indians. *Economic Botany* 38 (1): 52–64.

Dore, W. G. 1970. A Wild Ground-bean, *Amphicarpa,* for the Garden. *Greenhouse Garden Grass* 9 (2): 7–11.

Dorsey, J. Owen. 1881. Omaha Sociology. *Smithsonian Institution, Bureau of American Ethnology, 3rd Annual Report.*

———. 1894. Siouan Cults. *Smithsonian Institution, Bureau of American Ethnology, 11th Annual Report.*

Dunbar, John D. 1880. The Pawnee Indians. *Magazine of American History* 5 (5): 321–342.

Earle, F. R., and Quentin Jones. 1962. Analyses of Seed Samples from 113 Plant Families. *Economic Botany* 16 (4): 221–250.

Elias, Thomas S., and Peter A. Dykeman. 1982. *Field Guide to North American Edible Wild Plants.* New York: Van Nostrand Reinhold Co.

Elvin-Lewis, Memory. 1979. Empirical Rationale for Teeth Cleaning Plant Selection. *Medical Anthropology* Fall 3: 431–456.

Erichsen-Brown, Charlotte. 1979. *Use of Plants for the Past 500 Years.* Aurora, Ont.: Breezy Creeks Press.

Ewers, John D. 1958. *The Blackfeet: Raiders on the Northwestern Plains.* Norman: University of Oklahoma Press.

Felger, Richard S. 1979. Ancient Crops for the Twenty-first Century. In: *New Agricultural Crops.* Edited by Gary Richie. Boulder, Colo.: Westview Press.

Fenton, William N., editor. 1968. *Parker on the Iroquois.* Syracuse, N.Y.: Syracuse University Press.

Fernald, Merritt Lyndon. 1950. *Gray's Manual of Botany* (8th ed.). New York: American Book Co.

Fernald, Merritt Lyndon, Alred Charles Kinsey, and Reed C. Rollins. 1958. *Edible Wild Plants of Eastern North America.* New York: Harper and Row.

Fewkes, J. W. 1896. A Contribution to Ethnobotany. *American Anthropologist* (Series 1) 9: 14–21.

Fletcher, Alice C., and Francis La Flesche. 1911. The Omaha Tribe. *Smithsonian Institution, Bureau of American Ethnology,* 27th Annual Report.

Ford, Richard I. 1981a. Gardening and Farming before A.D. 1000: Patterns of Prehistoric Cultivation North of Mexico. *Journal of Ethnobiology* 1 (1): 6–27.

———. 1981b. Ethnobotany in North America: An Historical Phytogeographic Perspective. *Canadian Journal of Botany* 59: 2178–2189.

Forest Service, United States Department of Agriculture. 1937. *Range Plant Handbook.*

Foster, Steven. 1984. *Herbal Bounty! The Gentle Art of Herb*

Culture. Salt Lake City: Peregrine Smith Books.

Gaertner, Erika E. 1979. The History and Use of Milkweed (*Asclepias syriaca* L.). *Economic Botany* 33 (2): 119–123.

Gilmore, Melvin. 1913a. A Study in the Ethnobotany of the Omaha Indians. *Nebraska State Historical Society* 17: 314–357.

———. 1913b. Some Native Nebraska Plants with Their Uses by the Dakota. *Nebraska State Historical Society Proceedings and Collections* 17: 358–370.

———. 1913c. The Aboriginal Geography of the Nebraska Country. *Mississippi Valley Historical Association Proceedings* 6: 317–331.

———. 1914. Trip with White Eagle Determining Pawnee Sites. Nebraska State Historical Society Archives, Unpublished Manuscript No. 231.

———. 1921. The Ground Bean and the Bean Mouse and Their Economic Relations. *Annals of Iowa* (Series 3) 12: 606–609.

———. 1925. The Ground Bean and Its Uses. *Indian Notes* 2: 178–187.

———. 1926a. Arikara Commerce. *Indian Notes* 3: 13–18.

———. 1926b. Some Interesting Indian Foods. *Good Health* 61 (7): 12–14.

———. 1927. Indians and Conservation of Native Life. *Torreya* 27 (6): 97–98.

———. 1929. *Prairie Smoke.* New York: Columbia University Press.

———. 1931a. Dispersal by Indians, A Factor in the Extension of Discontinuous Distribution of Certain Species of Native Plants. *Michigan Academy of Science, Arts, and Letters* Paper 13: 89–94.

———. 1931b. Vegetal Remains of the Ozark Bluff-Dweller Culture. *Michigan Academy of Science, Arts, and Letters* Paper 14: 83–102.

———. 1977 (1919). *Uses of Plants by the Indians of the Missouri River Region.* Lincoln: University of Nebraska Press.

Gleason, Henry A., and Arthur Cronquist. 1963. *Manual of Vascular Plants of Northeastern United States and Adjacent Canada.* New York: Van Nostrand Reinhold Co.

Great Plains Flora Association. 1977. *Atlas of the Flora of the Great Plains.* Ames: Iowa State University Press.

———. 1986. *Flora of the Great Plains.* Lawrence: University Press of Kansas.

Gregg, Josiah. 1954. *Commerce of the Prairies.* Edited by Max L. Moorhead (originally published in 1844). Norman: University of Oklahoma Press.

Grinnell, George Bird. 1962. *The Cheyenne Indians.* Vol. 2. New York: Cooper Square Publishers.

Gudde, Erwin G., and E. K. Gudde, editors. 1958. *Exploring with Fremont: The Private Diaries of Charles Pruess.*

Norman: University of Oklahoma Press.

Harrington, H. D. 1967. *Edible Native Plants of the Rocky Mountains.* Albuquerque: University of New Mexico Press.

Hart, Jeffrey A. 1976. *Montana: Native Plants and Early Peoples.* Helena: Montana Historical Society.

———. 1981. The Ethnobotany of the Northern Cheyenne Indians of Montana. *Journal of Ethnopharmacology* 4: 1–55.

Havard, V. 1895. Food Plants of the North American Indians. *Bulletin of the Torrey Botanical Club* 22: 98–123.

———. 1896. Drink Plants of the North American Indians. *Bulletin of the Torrey Botanical Club* 23 (2): 33–46.

Hayden, F. V. 1859. Botany. In: *Report of the [U.S.] Secretary of War,* pp. 726–747.

———. 1862. On the Ethnology and Philology of the Indian Tribes of the Missouri Valley. *American Philosophical Society Transactions* 12 (2): 369–370.

Hedrick, U. P. 1919. *Sturtevant's Notes on Edible Plants.* 27th Annual Report, Vol. 2, Part II. New York Agricultural Experiment Station.

Heiser, Charles B., Jr. 1951. The Sunflower among the North American Indians. *Proceedings of the American Philosophical Society* 95 (4): 432–448.

———. 1985. Some Botanical Considerations of the Early Domesticated Plants North of Mexico. In: *Prehistoric Food Production in North America.* Edited by Richard I. Ford. Anthropological Paper 75, Museum of Anthropology, University of Michigan.

Hellson, John C. 1974. Ethnobotany of the Blackfoot Indians. *Canadian Ethnology Service Paper* 19, National Museum of Man, Ottawa.

Hinman, C. Wiley. 1984. New Crops for Arid Lands. *Science* 225: 1445–1448.

Hough, W. 1896. The Hopis' Relation to Their Plant Environment. *American Anthropologist* (Series 1) 10:33.

Howard, James H. 1965. The Ponca Tribe. *Smithsonian Institution, Bureau of American Ethnology. Bulletin* 195: 43–46.

Irving, John Treat, Jr. 1955. *Indian Sketches.* Edited by John F. McDermott (originally published 1833). Norman: University of Oklahoma Press.

Jackson, Donald, and Mary Lee Spence, editors. 1970. *The Expeditions of John Charles Frémont.* Vol. 1. Urbana: University of Illinois.

Jackson, Wes. 1985. *New Roots for Agriculture.* Lincoln: University of Nebraska Press.

Jackson, Wes, and Marty Bender. 1978. Eastern Gama Grass: Grain Crop for the Future? *Land Report* 6: 6–9.

Jenks, Albert E. 1900. The Wild Rice Gatherers of the Upper

Great Lakes. *Smithsonian Institution, Bureau of American Ethnology, 33rd Annual Report.*

Johnson, Alfred E. 1981. The Kansas City Hopewell Subsistence and Settlement System. *Missouri Archeologist* 42: 69–76.

Johnston, Alex. 1962. *Chenopodium album* as a Food Plant in Blackfoot Indian Prehistory. *Ecology* 43 (1): 129–130.

———. 1970. Blackfoot Indian Utilization of the Flora of the Northwestern Great Plains. *Economic Botany* 24: 301–324.

Kaldy, M. S., A. Johnston, and D. B. Wilson. 1980. Nutritive Value of Indian Bread-root, Squaw-root, and Jerusalem Artichoke. *Economic Botany* 34 (4): 352–357.

Kennedy, Michael Stephen, editor. 1961. *The Assiniboines—From the Accounts of the Old Ones Told to First Boy (James Larpenteur).* Norman: University of Oklahoma Press.

Kindscher, Kelly. 1982. *The Kansas Food System: Analysis and Action toward Sustainability.* Emmaus, Pa.: Rodale Press.

Kirk, Donald R. 1970. *Wild Edible Plants of the Western United States.* Heraldsburg, Calif.: Naturgraph Press.

Komarek, E. V. 1965. Fire Ecology—Grasslands and Man. *Proceedings of the 4th Annual Tall Timbers Fire Ecology Conference*, pp. 169–220.

Krochmal, A., S. Paur, and P. Duisberg. 1954. Useful Native Plants in the American Southwestern Deserts. *Economic Botany* 8: 3–20.

La Flesche, Francis. 1932. A Dictionary of the Osage Language. *Smithsonian Institution, Bureau of American Ethnology, Bulletin* 109.

Landes, Ruth. 1968. *The Mystic Lake Sioux: Sociology of the Mdewakantowan Santee.* Madison: University of Wisconsin Press.

Lehmer, Donald J. 1954. Archeological Investigations of the Oahe Dam Area, South Dakota. *Smithsonian Institution, Bureau of American Ethnology,* Bulletin 158, Appendix 1—*Vegetal Remains,* p. 163.

Lewis, Walter H., and Memory Elvin-Lewis. 1977. *Medical Botany.* New York: Wiley.

Maisch, John M. 1889. Useful Plants of the Genus Psoralea. *American Journal of Pharmacy* 61: 345–352.

Malin, James C. 1961. *The Grassland of North America: Prolegomena to its History.* Lawrence, Kans.: By the author.

Mandelbaum, David G. 1940. The Plains Cree. *American Museum of Natural History, Anthropology Paper* 37: 202–203.

Mattes, Merrill J. 1952. Capt. L. C. Easton's Report: Fort

Laramie to Fort Leavenworth via Republican River in 1849. *Kansas Historical Quarterly* 20: 392–415.

Matthews, John J. 1961. *The Osages, Children of the Middle Waters*. Norman: University of Oklahoma Press.

McClintock, Walter. 1909. Materia Medica of the Blackfeet. *Zeitschrift für Ethnologie* 51: 273–279.

McFarling, Lloyd. 1955. *Exploring the Northern Plains, 1804–1876*. Caldwell, Idaho: Caxton Printers.

McKelvey, Susan Delano. 1955. *Botanical Explorations of the Trans-Mississippi West*. Jamaica Plain, Mass.: Arnold Arboretum.

Medsger, Oliver Perry. 1966 (1939). *Edible Wild Plants*. New York: Collier-Macmillan Publishers.

Millspaugh, Charles F. 1974 (1892). *American Medicinal Plants*. New York: Dover.

Mooney, James. 1896. Calendar History of the Kiowa Indians. *Smithsonian Institution, Bureau of American Ethnology, 17th Annual Report*.

Moore, Michael. 1979. *Medicinal Plants of the Mountain West*. Santa Fe: Museum of New Mexico Press.

Morris, Elizabeth Ann, W. Max Witkind, Ralph L. Dix, and Judith Jacobson. 1981. Nutritional Content of Selected Aboriginal Foods in Northeastern Colorado: Buffalo (*Bison bison*) and Wild Onions (*Allium sp.*). *Journal of Ethnobiology* 1 (2): 213–220.

Morton, Julia F. 1963. Principal Wild Food Plants of the United States. *Economic Botany* 17 (4): 319–330.

Munson, Patrick J. 1981. Contributions to Osage and Lakota Ethnobotany. *Plains Anthropology* 26: 229–240.

Nabhan, Gary, Alfred Whiting, Henry Dobyns, Richard Hevly, and Robert Euler. 1981. Devil's Claw Domestication: Evidence from Southwestern Indian Fields. *Journal of Ethnobiology* 1 (1): 135–164.

Nabhan, Gary, and J. M. J. De Wet. 1984. *Panicum sonorum* in Sonoran Desert Agriculture. *Economic Botany* 38 (1): 65–82.

Nebraska Statewide Arboretum. No date. *Common & Scientific Names of Nebraska Plants: Native and Introduced*. Forestry Sciences Laboratory Publication 101. Lincoln: University of Nebraska.

Neihardt, John G. 1952. *When the Tree Flowered*. New York: Macmillan Company, pp. 152–170.

Newberry, J. S. 1888. Food and Fiber Plants of the North American Indians. *Popular Science Monthly* 32: 31–46.

Nickel, Robert K. 1974. *Plant Resource Utilization at a Late Prehistoric Site in North-central South Dakota*. Master's thesis, Department of Anthropology, University of Nebraska.

———. 1977. The Study of Archaeologically Recovered Plant Materials from the Middle Missouri Subarea. *Plains Anthropology* 22 (78, part 2): 53–58.

Nielsen, P. E. 1977. Plant Crops as a Source of Fuel and Hydrocarbon Like Materials. *Science* 198: 942–944.

Niethammer, Carolyn. 1974. *American Indian Food and Lore.* New York: Macmillan Publishing Co.

Nuttall, Thomas. 1980 (1821). *A Journal of the Travels into the Arkansas Territory during the Year 1819.* Edited by Savoie Lottinville. Norman: University of Oklahoma Press.

O'Brien, Patricia. 1982. The Yeo Site: A Kansas City Hopewell Limited Activity Site in Northwest Missouri, and Some Theories. *Plains Anthropologist* 27 (95): 37–56.

Opler, M. E. 1936. Introduction to Mescalero and Chiricahua Apache Cultures. *New Mexico University Biology Series Bulletin* 4 (5): 3–63.

Palmer, Edward. 1871. Food Products of the North American Indians. *Annual Report of the Commissioner of Agriculture,* House Executive Document, 3rd Session, 41st Congress, Serial Set No. 1461, pp. 404–428.

———. 1878. Plants Used by the Indians of the United States. *American Naturalist* 12: 593–

606 (Sept.) and 646–655 (Oct.).

Parsons, J. J. 1985. On "Bioregionalism" and "Watershed Consciousness." *Professional Geographer* 37 (1): 1–6.

Payne, Willard W., and Volney H. Jones. 1962. The Taxonomic Status and Archaeological Significance of a Giant Ragweed from Prehistoric Bluff Shelters in the Ozark Plateau Region. *Michigan Academy of Science, Arts, and Letters* Paper 47: 147–163.

Peterson, Lee. 1978. *A Field Guide to Edible Wild Plants of Eastern and Central North America.* Boston: Houghton Mifflin Co.

Phillips, Jan. 1979. *Wild Edibles of Missouri.* Jefferson City: Conservation Commission of the State of Missouri.

Phillips Petroleum Company. 1959. *Pasture and Range Plants.* Bartlesville, Okla.

Pinkerton, John. 1812. Kalm's Travels in North America. In: *Voyages and Travels in All Parts of the World.* London: Longman, Hurst et al.

Prescott, Philander. 1849. Farming among the Sioux Indians. *U.S. Patent Office Report on Agriculture,* pp. 451–455.

Reagan, Albert B. 1929. Plants Used by the White Mountain Apache Indians of Arizona. *Wisconsin Archeologist* 8: 143–161.

Reed, Flo. 1970 (reprint). *Uses of*

Native Plants by Nevada Indians. Carson City: State of Nevada, Department of Education.

Reichart, Milton. 1983. Psoralea Esculenta in Northeast Kansas. *Journal of the Kansas Anthropological Association* 4: 103–121.

Reid, Kenneth C. 1977. *Psoralea esculenta* as a Prairie Resource: An Ethnographic Appraisal. *Plains Anthropologist* 22 (78, part 1): 321–327.

Robbins, Wilfred, John P. Harrington, and Barbara Freire-Marreco. 1916. Ethnobotany of the Tewa. *Smithsonian Institution, Bureau of American Ethnology, Bulletin* 55.

Robertson, J. H. 1939. Study of True-Prairie Vegetation. *Ecological Monographs* 9 (4): 458–461.

Rock, Harold W. 1977. *Prairie Propagation Handbook.* Milwaukee: Milwaukee County Park System, Wehr Nature Center.

Rogers, Dilwyn J. 1980a. *Edible, Medicinal, Useful, and Poisonous Wild Plants of the Northern Great Plains–South Dakota Region.* Sioux Falls, S.D.: Biology Department, Augustana College.

———. 1980b. *Lakota Names and Traditional Uses of Native Plants By Sicangu (Brule) People in the Rosebud Area, South Dakota.* St. Francis, S.D.: Rosebud Educational Society.

Rollins, Reed C. 1939. The Cruciferous Genus Stanleya. *Lloydia* 12 (2): 109–118.

Rusby, H. H. 1906a. The August Wild Foods of the United States. *Country Life in America* August: 437, 446.

———. 1906b. More June Wild Foods. *Country Life in America* June: 220.

———. 1907. Wild Foods of the United States in May. *Country Life in America* March: 66–69.

Salac, S. S., P. N. Jensen, J. A. Dickerson, and R. W. Gray, Jr. 1978. Wildflowers for Nebraska Landscapes. *Nebraska Agricultural Experiment Station,* MP 35.

Sale, Kirkpatrick. 1985. *Dwellers in the Land: The Bioregional Vision.* San Francisco: Sierra Club Books.

Sheldon, Addison E. 1923. A Nebraska Bean—Worthy Cultivation. *Nebraska History* 6 (3): 79.

Smith, Huron H. 1928. Ethnobotany of the Meskwaki. *Bulletin of the Public Museum of the City of Milwaukee* 4 (2): 175–326.

———. 1933. Ethobotany of the Forest Potawatomi. *Bulletin of the Public Museum of the City of Milwaukee* 7 (1): 105–127.

Smith, J. Robert, and Beatrice S. Smith. 1980. *The Prairie Garden.* Madison: University of Wisconsin Press.

Stephens, Homer A. 1980.

Poisonous Plants of the Central United States. Lawrence: Regents Press of Kansas.

Stevens, William Chase. 1961. *Kansas Wild Flowers.* Lawrence: University of Kansas Press.

Stevenson, Matilda. 1915. Ethnobotany of the Zuni. *Smithsonian Institution, Bureau of American Ethnology, 30th Annual Report.*

Steyermark, Julian A. 1981 (1963). *Flora of Missouri.* Ames: Iowa State University Press.

Thwaites, Reuben Gold, editor. 1904. *Original Journals of the Lewis and Clark Expedition.* New York: Dodd, Mead and Co. 6 Vols.

———. 1905, 1906. *Early Western Travels.* Cleveland: Arthur Clark Company.

Vol. 6. Brackenridge's Journal up the Missouri, 1811.

Vol. 14, 15, 16. Edwin James' Account of the Steven H. Long Expedition, 1819–20.

Vol. 23, 24. Prince Maximilian of Wied's Travels, 1832–4.

Vol. 26. The Far West; or, a Tour beyond the Mountains, by Edmund Flagg, 1836–41.

Vol. 27. Letters and Sketches, with a Narrative of a Year's Residence among the Indian Tribes of the Rocky Mountains, by Father Pierre J. de Smet, 1836–41.

Vol. 28. Travels in the Great Western Prairies, the Anahuac and Rocky Mountains, and in the Oregon Territory, by

Thomas J. Farnham, 1839–46.

Timbrook, Jan. 1982. Use of Wild Cherry Pits as Food by the California Indians. *Journal of Ethnobiology* 2 (2): 162–176.

Trumbull, J. Hammond and Asa Gray. 1877. Notes on the History of Helianthus tuberosus, the so called Jerusalem Artichoke. *American Journal of Science and Arts* (Series III) 13: 347–352.

Turner, Nancy J. 1981. A Gift for the Taking: The Untapped Potential of Some Food Plants of North American Native Peoples. *Canadian Journal of Botany* 59 (11): 2321–2357.

Vestal, Paul A., and Richard Evans Schultes. 1939. *The Economic Botany of the Kiowa Indians.* Cambridge, Mass.: Botanical Museum.

Vietmeyer, Noel D. 1981. Rediscovering America's Forgotten Crops. *National Geographic* 159 (5): 709–712.

Wakefield, E. G., and Samuel C. Dellinger. 1936. Diet of Bluff Dwellers of the Ozark Mountains and its Skeletal Effects. *Annals of Internal Medicine* 9: 1412–1418.

Wallace, Ernest, and E. Adamson Hoebel. 1972. *The Comanches—Lords of the South Plains.* Norman: University of Oklahoma Press.

Walter, William, Edward Croom, George Catignani, and Wayne Thresher. 1986. Compositional Study of *Apios priceana* Tubers. *Journal of Agricultural and*

Food Chemistry 34 (1): 39–41.

Watt, Bernice K., and Annabel L. Merrill. 1963. *Composition of Foods.* Agricultural Handbook 8, U.S. Department of Agriculture.

Weaver, John Ernest. 1919. *The Ecological Relations of Roots.* Carnegie Institute of Washington Publication 286.

———. 1968. *Prairie Plants and Their Environment.* Lincoln: University of Nebraska Press.

Weaver, John Ernest, and T. J. Fitzpatrick. 1934. The Prairie. *Ecological Monographs* 4: 109–295 (reprinted by the Prairie-Plains Resource Institute, Aurora, Nebr., 1980).

Wedel, Waldo R. 1936. An Introduction to Pawnee Archeology. *Smithsonian Institution, Bureau of American Ethnology, Bulletin* 112.

———. 1955. Archeological Materials from the Vicinity of Mobridge, South Dakota. *Smithsonian Institute, Bureau of Ethnology, Bulletin* 157 (Anthropology Paper 45), pp. 144–146.

———. 1978. Notes on the Prairie Turnip (Psoralea esculenta) among the Plains Indians. *Nebraska History* 59 (2): 1–25.

Weitzner, Bella. 1979. Notes on the Hidatsa Indians Based on Data Recorded by the Late Gilbert L. Wilson. *Anthropological Papers of the American Museum of Natural History* 56 (2): 212–217.

Whiting, Alfred, E. 1939. The Ethnobotany of the Hopi. *Museum of Northern Arizona, Bulletin* 15.

Whitney, Chauncey B. 1911. Diary of Chauncey B. Whitney. *Kansas State Historical Society* 12: 296–299.

Williams-Dean, Glenna Joyce. 1978. *Ethnobotany and Cultural Ecology of Prehistoric Man in Southwest Texas.* Ph.D. dissertation, Texas A & M University.

Wilson, Gilbert L. 1981 (1927). *Waheenee: An Indian Girl's Story.* Lincoln: University of Nebraska Press.

Winter, Joseph Charles. 1974. *Aboriginal Agriculture in the Southwest and Great Basin.* Ph.D. dissertation, University of Utah.

Winter, Joseph C., and Henry G. Wylie. 1974. Paleoecology and Diet at Clydes Cavern. *American Antiquity* 39 (2): 303–315.

Wissler, Clark. 1910. Material Culture of the Blackfoot. *American Museum of Natural History Anthropological Papers* 5: 22–23.

Witthoft, John. 1977. Cherokee Use of Potherbs. *Journal of Cherokee Studies* 2 (2): 250–255.

Wyman, Leland C., and Stewart K. Harris. 1951. *Ethnobotany of the Kayenta Navaho.* Albuquerque: University of New Mexico Press.

Yanovsky, Elias. 1936. Food Plants

of the North American Indians. *Miscellaneous Publication* 237, U.S. Department of Agriculture.

Yanovsky, E., and R. M. Kingsbury. 1938. Analyses of Some Indian Food Plants. *Association of Official Agricultural Chemists* 21 (4): 648–655.

Yarnell, Richard A. 1971. Early Woodland Plant Remains and the Question of Cultivation. In: *Prehistoric Agriculture.* Edited by Stuart Struever. Garden City, N.Y.: Natural History Press, pp. 550–554.

Yarnell, Richard A. 1972. Iva annua var. macrocarpa: Extinct American Cultigen? *American Anthropologist* 74 (3): 335–341.

———. 1978. Domestication of Sunflower and Sumpweed in Eastern North America. In: *The Nature and Status of Ethnobotany.* Edited by Richard I. Ford. Anthropological Paper 67, Museum of Anthropology, University of Michigan, pp. 289–299.

Zennie, Thomas M., and C. Dwayne Ogzewalla. 1977. Ascorbic Acid and Vitamin A Content of Edible Wild Plants of Ohio and Kentucky. *Economic Botany* 31: 76–79.

Index

225, 236, 243, 244, 246; plant use: American lotus, 245–46; beebalm, 151; buffalo berry, 211;buffalo gourd, 107; calamus, 238; chokecherry, 180; common milkweed, 56; elderberry, 250; groundnut, 48, 50; groundplum milkvetch, 62; hazelnut, 99; hoary peavine, 243; hog peanut, 38–39; Jerusalem artichokes, 130–31; lamb's quarters, 81; prairie turnip, 186; serviceberry, 30; violet, 221; wild onions, 14; wild plums, 171; wild rice, 236–37; wild rose, 201; wood sorrel, 160
Opuntia ficus-indica, 157
Opuntia macrorhiza, 153–57
Opuntia phaeacantha, 157
Opuntia polyacantha, 154
Orach, 66–67
Oregon Trail, 70
Orobanche ludoviciana, 247–48
Oryzopsis hymenoides, 232–33
Osage Indians, 5; names for plants, 13, 106, 116, 177, 184, 238; plant use: calamus, 238; chokecherry, 180; gayfeather, 144; groundnut, 49; hackberry, 242; purple poppy mallow, 69; wild rose, 201; wood sorrel, 160
Otoe Indians, 245
Oxalis stricta, 159–60
Oxalis tuberosa, 160
Oxalis violacea, 158–60
Oxytropis lambertii, 62
Oxytropis sericea, 62
Ozark Bluff-dweller Indians, 6; plant use: calamus, 238; gayfeather, 144; giant ragweed, 25; groundnut, 49; lamb's quarters, 82–83; pigweed, 22; soapweed, 226; sumac, 192; wild onions, 15

Paiute Indians, 6; cultivation of pigweed, 22; plant use: American licorice, 121; bastard toadflax, 97; prince's plume, 216
Palmer, Edward: cultivation of pig-

weed, 22; on American licorice, 121; on bastard toadflax, 97; on broomrape, 248; on bush morning glory, 135–36; on common milkweed, 56; on devil's claw, 167; on dewberry and blackberry, 207; on groundnut, 50; on Indian ricegrass, 233; on lamb's quarters, 81; on prickly pear, 155; on prince's plume, 216; on sunflower, 125
Panamint (Koso) Indians, 233
Panic grass, 233
Panicum bulbosum, 233
Panicum capillare, 233
Panicum obtusum, 233
Panicum sonorum, 233
Panicum virgatum, 3
Pansy violet, 223
Papago Indians, 168
Parasitic plants, 97
Passenger pigeons, 39
Pawnee Indians, 5; names for plants, 13, 38, 47, 55, 110, 116, 120, 124, 130, 135, 143, 150, 159, 162, 170, 177, 191, 200, 211, 225, 238, 242, 244; plant use: American licorice, 121; American lotus, 245; beebalm, 151; buffalo berry, 211; bush morning glory, 135; calamus, 238; chokecherry, 180; common milkweed, 56; elderberry, 250; groundnut, 49; hog peanut, 39; Jerusalem artichoke, 130; lamb's quarters, 81; prickly pear, 155; wild onions, 14; wild plums, 174; wild rose, 201, 204; wild strawberry, 117; wood sorrel, 160
Pemmican, made with: buffalo berry, 212; chokecherry, 178–80; currants, 197; serviceberry, 29–32
Peppermint, 178
Perennial grain crops, 229, 235–36
Phalaris caroliniana, 25, 233–34, 248
Phillips Spring, 163
Physalis heterophylla, 161
Physalis lobata, 163
Physalis longifolia, 164

Pigweed, 18–22, 25
Pike, Zebulon, 49
Pima Indians, 6; plant use: devil's claw, 168; prickly pear, 155; saltbush, 65–66
Pincushion cactus, 102–4
Pioneers, 5
Plains Cree Indians, 5; cultivation of chokecherry, 182; names for plants, 28, 154; plant use: dewberry and blackberry, 207; lavender hyssop, 239; prairie turnip, 185, 187; serviceberry, 30
Plums. *See* Wild plums
Poaceae, 228–37
Poisonous plants: avoidance of, 8; chokecherry, 181; death camas, 16, 73, 74–75; dogbane, 58; elderberry, 250; ground cherry, 163; hoary peavine, 243; locoweed, 62; milkweeds, 58; poisonous milkvetch, 62–63; prince's plume, 216; saltbush, 66;
Polygonum amphibium, 248
Polygonum bistortoides, 47
Polygonum coccineum, 248
Polygonum erectum, 248
Pomme de terre, 47, 50
Ponca Indians, 5; names for plants, 13, 28, 33, 38, 47, 55, 61, 99, 106, 110, 116, 124, 143, 150, 159, 162, 170, 184, 191, 196, 200, 210, 225, 236, 243, 244; plant use: beebalm, 151; buffalo berry, 211; buffalo gourd, 107; elderberry, 250; groundplum milkvetch, 62; hazelnut, 99; hoary peavine, 243; hog peanut, 39; Jerusalem artichoke, 130; prairie clovers, 111; serviceberry, 30; wild onions, 14; wild rose, 201; wild strawberry, 117
Poppy mallow, 68–71
Potawatomie Indians, 6; cultivation of common milkweed, 58; use of hazelnut, 100
Potentilla species, 248–49
Prairie Bioregion, 2–5

Prairie clover, 109–11
Prairie dogs, 21
Prairie fawnlily, 113–14
Prairie fires, effect of, 1–2, 100
Prairie parsley, 146–48
Prairie turnip, 4, 29, 30, 41, 47, 50, 70, 183–89, 211
Prairie violet, 220–23
Prairie wild rose, 199–204
Prescott, Philander, 48
Prickly pear, 153–57
Prince's plume, 215–16
Proboscidea fragrans, 167
Proboscidea louisianica, 165–68
Proboscidea parviflora, 168
Prostrate pigweed, 18–22
Pruess, Charles, 3
Prunus americana, 5, 117, 169–75, 197, 207, 212
Prunus angustifolia, 171, 173, 175
Prunus gracilis, 171
Prunus ilicifolia, 181
Prunus nigra, 174
Prunus pumila, 170, 173–74
Prunus serotina, 170
Prunus virginiana, 4, 5, 30, 170, 176–82, 197, 211, 212
Psoralea cuspidata, 188
Psoralea esculenta, 4, 29, 30, 41, 47, 50, 70, 183–89, 211
Psoralea hypogaea, 188
Pueblo Indians, 6; plant use: beebalm, 151; greenthread, 251; lamb's quarters, 81; Rocky Mountain bee plant, 93; pigweed, 20; saltbush, 66. *See also specific tribes*
Pumpkins, cultivated by the Tewa, 93
Purple poppy mallow, 68–71
Purple prairie clover, 110–11
Pussy-toes, 239–40
Pycnanthemum species, 249

Rafinesque, Constantine, 51
Raspberry, 100
Rhus aromatica, 191–94
Rhus glabra, 190–94

Library of Congress Cataloging-in-
Publication Data
Kindscher, Kelly.
  Edible wild plants of the prairie.
  Bibliography: p.
  Includes index.
  1. Wild plants, Edible—Great Plains.
2. Wild plants, Edible—Middle West.
3. Prairie flora—Great Plains. 4. Prairie
flora—Middle West. 5. Indians of
North America—Food. 6. Indians of
North America—Ethnobotany. 7. Eth-
nobotany—Great Plains. 8. Ethnobo-
tany—Middle West. I. Title.
QK98.5.U6K56 1987   581.6′32′0978
87-6162
ISBN 0-7006-0324-7
ISBN 0-7006-0325-5 (pbk.)